Biomimetic, Bio-inspired and Self-Assembled Materials for Engineered Surfaces and Applications

T0351301

MATERIALS RESEARCH SOCIETY
SYMPOSIUM PROCEEDINGS VOLUME 1498

Biomimetic, Bio-inspired and Self-Assembled Materials for Engineered Surfaces and Applications

Symposium held November 25–30, 2012, Boston, Massachusetts, U.S.A.

EDITORS

Michelle L. Oyen

Cambridge University
Cambridge, United Kingdom

Shelly R. Peyton

University of Massachusetts, Amherst
Amherst, Massachusetts, U.S.A.

Gila E. Stein

University of Houston
Houston, Texas, U.S.A.

Materials Research Society
Warrendale, Pennsylvania

CAMBRIDGE
UNIVERSITY PRESS

Shaftesbury Road, Cambridge CB2 8EA, United Kingdom

One Liberty Plaza, 20th Floor, New York, NY 10006, USA

477 Williamstown Road, Port Melbourne, VIC 3207, Australia

314–321, 3rd Floor, Plot 3, Splendor Forum, Jasola District Centre, New Delhi – 110025, India

103 Penang Road, #05–06/07, Visioncrest Commercial, Singapore 238467

Cambridge University Press is part of Cambridge University Press & Assessment, a department of the University of Cambridge.

We share the University's mission to contribute to society through the pursuit of education, learning and research at the highest international levels of excellence.

www.cambridge.org
Information on this title: www.cambridge.org/9781605114750

Materials Research Society
506 Keystone Drive, Warrendale, PA 15086
http://www.mrs.org

First published 2013

Army Research Office (ARO) support was provided under Grant W911NF-13-0021. The views, opinions, and/or findings contained in this report are those of the author(s) and should not be construed as an official Department of the Army position, policy, or decision, unless so designated by other documentation.

CODEN: MRSPDH

A catalogue record for this publication is available from the British Library

ISBN   978-1-605-11475-0   Hardback

CONTENTS

BIOINSPIRED DIRECTIONAL SURFACES–FROM NATURE TO ENGINEERED TEXTURED SURFACES

FUNCTIONAL AND RESPONSIVE MATERIALS EXPLOITING PEPTIDE AND PROTEIN SELF-ASSEMBLY

*Invited Paper

FUNDAMENTALS OF ASSEMBLY IN BIOMOLECULAR AND BIOMIMETIC SYSTEMS

DIRECTED SELF-ASSEMBLY FOR NANOPATTERNING

PREFACE

This volume represents a collection of papers from five distinct yet related symposia: Symposium L, "Biomimetic Nanoscale Platforms, Particles, and Scaffolds for Biomedical Applications"; Symposium M, "Bioinspired Directional Surfaces–From Nature to Engineered Textured Surfaces"; Symposium Q, "Functional and Responsive Materials Exploiting Peptide and Protein Self-Assembly"; Symposium R, "Fundamentals of Assembly in Biomolecular and Biomimetic Systems"; and Symposium S, "Directed Self-Assembly for Nanopatterning." This volume is thus the first for which the papers are from a larger group of thematically-related symposia, with emphasis on bioinspiration and biomimicry, self-assembly and natural materials. There are 40 Proceedings papers contained herein, one from an invited speaker and 39 from contributed speakers, capturing a subset of the five individual symposia, in which a total of more than 500 papers were presented. The editors of this volume are indebted to the organizers of all five individual symposia for their efforts.

Michelle L. Oyen
Shelly R. Peyton
Gila E. Stein

May 2013

MATERIALS RESEARCH SOCIETY SYMPOSIUM PROCEEDINGS

MATERIALS RESEARCH SOCIETY SYMPOSIUM PROCEEDINGS

MATERIALS RESEARCH SOCIETY SYMPOSIUM PROCEEDINGS

Volume 1534E — Low-Dimensional Semiconductor Structures, 2012, T. Torchyn, Y. Vorobie, Z. Horvath, ISBN 978-1-60511-511-5

Prior Materials Research Society Symposium Proceedings available by contacting Materials Research Society

Biomimetic Nanoscale Platforms, Particles, and Scaffolds for Biomedical Applications

Mater. Res. Soc. Symp. Proc. Vol. 1498 © 2013 Materials Research Society
DOI: 10.1557/opl.2013.411

Improvement of antisense oligonucleotides delivery using high hydrostatic pressurized lipoplex

Tsuyoshi Kimura[1], Asami Sano[1], Kwangwoo Nam[1], Kazunari Akiyoshi[2], Yoshihiro Sasaki[1], Akio Kishida[1]

[1]Institute of Biomaterials and Bioengineering, Tokyo Medical and Dental University, 2-3-10 Kanda-surugadai, Chiyoda-ku Tokyo 101-0062, Japan
[2]Department of Polymer Chemistry, Kyoto University, Kyoto daigaku-Katsura, Nishikyo-ku, Kyoto 615-8530

ABSTRACT

Cationic liposome (CL) is a promising vector for nucleic acid therapy. In the present study, we investigated the effect of high hydrostatic pressure (HHP) treatment to lipoplex on the lipoplex-based antisense oligodeoxynucleotides (AS-ODNs) delivery in order to improve the transfection efficacy of lipoplex. Cationic liposome consisting of DOTMA and DOPE was used. AS-ODNs were designed to inhibit the expression of firefly luciferase. The complexes of CL and AS-ODN were prepared at various C/A ratios and then pressurized hydrostatically at various atmospheres (~10,000 atm) for 10 min (HHP treatment). After removal of pressure, the pressurized lipoplexes were used. The lipoplex with and without the HHP treatment was transferred into HeLa cells expressing firefly luciferase transiently. The luciferase activity using the HHP-treated lipoplex was decreased compared to that of the non-pressurized lipoplex. Also, for HEK293 cells expressing luciferase stably, the lipoplex with the HHP treatment could effectively suppress the luciferase expression. In order to elucidate relationship between the structure and the transfection efficiency of the HHP-treated lipoplex, the properties of the HHP-treated lipoplex were examined by various physicochemical analyses. The different physicochemical properties between the lipoplexes with and without HHP treatment were showed, suggesting that the nature of lipoplex was changed by the HHP treatment. We believe that this change of lipoplex properties by the HHP treatment affected the efficiency of gene suppression. This HHP treatment for lipoplex appears to be a promising contribution to gene and oligonucleotide delivery.

INTRODUCTION

As a promising strategy for the treatment of many intractable diseases, gene therapy, which is defined as the genetic modification for therapeutic benefit, has attracted many researchers. For gene therapy, it is required to delivery exogenous gene drugs, such as plasmid DNA and oligonucleotide, into cells healthy. So far, the delivery of plasmid DNA has been attempted using many kinds of gene vectors in order to replace a missing gene. Recently, modulation of gene expression using short nucleic acids, such as antisense oligodeoxynucleotide (AS-ODN) and small interfering RNA (siRNA), is one of attractive methods for gene therapy [1, 2]. Various types of gene, which cause disease, have been target of short nucleic acid based therapy. For this therapy, it is important to deliver nucleic acids into cells effectively and to inhibit expression of target gene selectively [3]. In many cases, cationic liposomes (CLs) are used as transfection reagents for nucleic acid delivery because stable and small complexes of cationic liposome and nucleic acids, called "lipoplex", are spontaneously formed by electrostatic interaction [4]. So far, many lipids modified chemically [5] and biologically [6] have been synthesized and attempted as transfection reagents. However, despite the research on lipoplex-mediated nucleic acid such as plasmid DNA, ODN and siRNA, the efficiency of lipoplex-mediated delivery system is lower than desirable. Also, effort to understanding the structure-activity relationship of lipoplexs in nucleic acids delivery is in progress in order to optimize lipoplex performance [7]. Some researchers reported that the structure of lipoplex affected the transfection efficiency [8, 9].

It is well known that pressure, which is one of intensive variables in thermodynamic as well as temperature and concentration, strongly affects on protein structure because the related weak interactions, such as electrostatic, hydrophobic and hydrogen bonding interactions, to protein structure formation are depended on pressure. Previously, it was reported that the phase transition of lipid bilayer was induced under high hydrostatic pressure condition. Also, we reported that high hydrostatic pressurization (HHP) treatment for the complex of CL and plasmid DNA could enhance the transfection efficiency [10]. In the present study, we investigated the effect of high hydrostatic pressure (HHP) treatment to lipoplex on the lipoplex-based AS-ODN delivery in order to improve the transfection efficacy of lipoplex.

EXPERIMENT

Liposome preparation: Liposome was prepared according to the extrusion method [11]. Briefly, N-[1-(2,3,-dioleyloxy)propyl]-N,N,N-trimethylammoniumchloride (DOTMA) and dioleoyl phosphatidyl ethanolamine were purchased from Avanti Polar Lipids (Alabaster, AL). They were solved in chloroform at the concentration of 10 mM, respectively. They were mixed in equal amount and the solvent was removed by argon gas flashing in order to form thin film on the sample bottle. The obtained thin film was dried under the reduced pressure for 2 hours to remove solvent completely. The lipids were hydrated using voltex in 1/10 PBS and through by extrusion with 100nm pore filter.

Antisense oligodeoxynucleotide design: Figure 1 shows the position of AS-ODNs, which were designed to inhibit the expression of luciferase 2 gene.

Preparation of HHP-treated lipoplex: The complex of cationic liposome and AS-ODNs were prepared at various ratios of cationic group of cationic liposome and anionic group of AS-ODNs. Lipoplex-solutions were injected each 50 μL to well of 384 well plate, respectively. These plates were sealed and put into the small bag filled with water. The bag was sealed up without air. Then, it was set into the high hydrostatic pressurization device (Dr.CHEF, (Kobe Steel, Ltd.)). The HHP treatment was carried out at 10,000 atm and 40 °C for 15 min. After removal of pressure, the pressurized lipoplexes were used.

Transfection Procedure: HeLa cell was used. Plasmid DNAs encoding firefly luciferase (pGL4.13, Promega) and renilla luciferase (pGL4.73, Promega) were used, respectively. The firefly luciferase gene is target of AS-ODN and the renilla luciferase gene is control gene. These plasmid DNAs were transfected into HeLa and CHO cells using Lipofectamin2000. After 24 hours incubation of transfection, AS-ODNs were transfected into their cells using the lipoplexes (DOTMA and DOPE) with and without HHP treatment (Figure 2). HEK293 cells expressing firefly luciferase stably was also used.

4

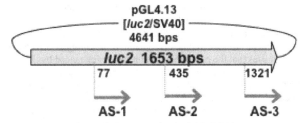

Figure 1. Target regions of AS-ODN.

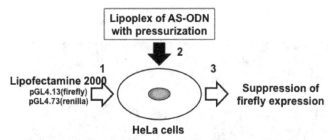

Figure 2. Scheme of transfection using the HHP-treated lipoplex.

RESULTS AND DISSCUSION

Several sequences of AS-ODN suppressing firefly luciferase expression were designed. They were mixed with CL at various C/A ratios and transferred into HeLa cells, in which plasmid DNAs encoding firefly and renilla luciferase were transiently transfected before the AS-ODN delivery. The activity of firefly luciferase was decreased, irrespective of sequences of the transfected AS-ODNs although the degree of luciferase activity depended on the used AS-ODNs (Table 1). Among them, the AS-2 AS-ODN showed the highest suppression of the luciferase expression without decreasing renilla luciferase activity.

Table 1. Gene suppression by AS-ODNs delivery into HeLa cells expressing firefly and renilla luciferase.

	Relative to control		
Samples	Non-pressurized lipoplex		HHP-treated lipoplex
	Firefly luciferase	Renilla luciferase	Firefly luciferase
AS-1	0.8 ±0.04	0.9 ±0.18	------
AS-2	0.44±0.09	0.85±0.07	0.31±0.03
AS-3	0.78±0.05	0.92±0.07	------

The lipoplex of AS-2 were prepared and pressurized at 10,000 atm and 40 °C for 10 min. When the lipoplexes with and without the HHP treatment were transfected into HeLa cells expressing firefly luciferase and renilla luciferase transiently, the firefly luciferase activity was effective decreased for the HHP-treated lipoplex compared to the non-pressurized lipoplex (Table 1). This result indicates that the expression inhibition was enhanced by the HHP-treated lipoplex. Also, for HEK293 cells expressing firefly luciferase stably, the effect of the HHP treatment on the lipoplex-based AS-ODN delivery was investigated. The level of gene suppression was decreased compared to that of HeLa cells expressing luciferase transiently. The efficiency of gene silencing was enhanced by the HHP-treated lipoplex compared to the non-pressurized lipoplex.

The cellular uptake of fluorescent-labeled AS-2 using the lipoplexes with and without HHP treatment was examined by confocal laser microscope (Figure 3). Many green fluorescent spots were observed for the HHP-treated lipoplex compared to the non-pressurized lipoplex, suggesting the effective cellular uptake of the HHP-treated lipoplex.

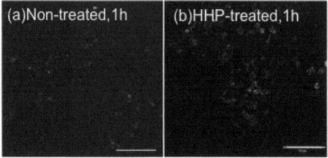

Figure 3. Confocal laser microscopic images of FITC labeled AS-ODN transfected by (a) the non-pressurized lipoplex and (b) HHP-treated lipoplex. Scale bar shows 50 μm.

It is well known that the properties of lipoplex, such as morphology and structure, affect on the transfection efficiency for nucleic acid delivery. In order to clarify the different transfection efficiency of lipoplex with and without the HHP treatment, the morphology and structure of them was investigated by atomic force microscope (AFM) observation, transmission electron microscope (TEM) observation, and fluorescent spectral analysis.

For AFM observation, spherical shape of the non-treated lipoplexes with the diameter of 100-200nm was observed, whereas bell-shaped form of the HHP-lipoplexes was observed with slight increasing of the diameter (Fig 4). The slight increasing of the diameter of the HHP treated lipoplex was confirmed by dynamic light scattering measurement (data not shown). It is suggested that the assembling of cationic liposome and AS-ODN was increased during the HHP treatment.

For TEM observation, the lamellar structure of the non-treated lipoplexes was observed, whereas the amorphous structure including small and regular structure was observed for the HHP-treated lipoplexes. These results suggest that the structure of the lipoplex was disordered by the HHP treatment. The structural transition may be induced by the HHP treatment.

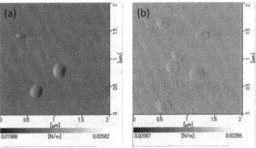

Figure 4. AFM observation of (a) the non-pressurized lipoplex and (b) HHP-treated lipoplex.

In order to examine the structural change of the HHP treated lipoplex, the hydrophobic microenvironment of lipoplex was investigated by using fluorescent hydrophobic probes, such as Laurdan and pyrene, which are membrane fluidity and hydrophobicity markers. Using Laurdan, generalized polarization (GP), which is sensitive to membrane phase transitions and other alterations to membrane fluidity, was calculated from the emission intensities using the following equation. GP=(I_{440}-I_{490})/ (I_{440}+I_{490}). The values of GP of the non-pressurized lipoplex and the HHP treated lipoplex 0.13 ± 0.04 and 0.07 ± 0.01. This suggests that the membrane fluidity of the HHP treated lipoplex was decreased because of the membrane fusion during the HHP treatment.

CONCLUSION

We successfully improved the efficiency of gene suppression by using the HHP-treated lipoplex. It was found that the change of the structure of lipoplex was induced by the HHP treatment. We believe that this change of lipoplex structure by the HHP treatment affected the efficiency of gene suppression. This HHP treatment for lipoplex appears to be a promising contribution to gene and ON delivery.

ACKNOWLEDGMENTS

This work was partly supported by grants from the Ministry of Health, Labor and Welfare, the Ministry of Education, Culture, Sports, Science and Technology, and Core Research for Evolutional Science and Technology (CREST) of the Japan Science and Technology Agency (JST).

REFERENCES

1.   Boussif, O., Lezoualch, F., Zanta, MA. et al., A versatile vector for gene and oligonucleotide transfer into cells in culture and in vivo: polyethylenimine, Proc. Nat. Acad. Sci., 1995, 92, 7297-7301.

2.   Zabner, J., Fasbender, AJ., Moninger, T. et al., Cellular and molecular barriers to gene transfer by a cationic lipid, J. Biol. Chem., 1995, 270, 18997-19007.
3.   Juliano, R., Bauman, J., Kang, H, Ming, X., Biological barriers to therapy with antisense and siRNA oligonucleotides, Mol. Pharm., 2009, 6(3), 686-695.
4.   Bilalov, A., Olsson, U., Lingman, B., Complexation between DNA and surfactants and lipids: phase behavior and molecular organization, Soft Matter, 2012, 8(43), 11022-11033.
5.   Midoux, P., Pichon, C., Yaouanc, JJ. Jaffres, PA., Chemical vectors for gene delivery: a current review on polymers, peptides and lipids containing histidine or imidazole as nucleic acids carriers, Br. J. Pharmacol. 2009, 157(2), 166-178.
6.   Yu, W., Zhang, N., Li, C., Saccharide modified pharmaceutical nanocarriers for targeted drug and gene delivery, Curr. Pharm. Des., 2009, 15(32), 3826–3836.
7.   Resina, S., Prevot, P., Therry, AR., Physico-chemical characteristics of lipoplexes influence cell uptake mechanisms and transfection efficacy, Plos One, 2009, 4(6) e6058.
8.   Sakuragi M., Kusuki, S., Hamada, E., et al. Supramolecular structures of benzyl amine derivate/DNA complexes explored with synchrotron small angle X-ray scattering at SPring-8, J. Physics: Conference Series, 2009, 184, 012008.
9.   Ma, B., Zhang, S., Jiang, H., Zhao, B., Lv, H., Lipoplex morphologies and their influences on transfection efficiency in gene delivery, J. Control. Release, 2007, 123, 184-194.
10.   Kimura, T., Sano, A., Kishida, A. et al. Gene transfection using pressurized lipoplex prepared by different pressurization procedure, 35th Annual Meeting & Exposition of the Controlled Release Society.
11.   Malaekeh-Nikouei, B., Malaekeh-Nikouei, M., Oskuee, RK., Ramezani, M., Preparation, characterization, transfection efficiency, and cytotoxicity of liposomes containing oligoamine-modified cholesterols as nanocarriers to Neuro2A cells, Nanomedicine, 2009, 5(4), 457-462.

Mater. Res. Soc. Symp. Proc. Vol. 1498 © 2012 Materials Research Society
DOI: 10.1557/opl.2012.1658

Antitumoral drug loaded in TEOS nanoparticles

Ana Paula V. Araújo[1]; Claure N. Lunardi[1], Anderson J. Gomes[1]

[1]University of Brasília - Faculdade de Ceilândia (UnB-FCE), Brasília, DF 7220-140, Brasil

ABSTRACT

Methotrexate (MTX), is a potent immunomodulating drug and widely used in the treatment of cancer, psoriasis and others disease. Despite its efficacy, the use of MTX is greatly limited due to its toxicity. To solve this problem, we prepared nanoparticles of tetraethyl orthosilicate (NP-TEOS) containing the compound methotrexate (MTX), by the sol-gel method. This drug delivery system (DDS) showed a loading efficiency of 39.7%. Size distribution studies were performed with dynamic light scattering and scanning electron microscopy revealing that these particles were spherical in shape, with a mean diameter between 140-430 nm and a low polydispersity (0.12 – 0.26). Also the particles displayed a low tendency toward aggregation which was confirmed by the low zeta potential -61.4 mV. Profile release showed a slow release loaded with MTX (PBS buffer pH = 7.4). The slow release can be attributed to the low porosity of the NP-TEOS and the extremely low diffusivity of MTX in aqueous media. B16-F10 cells were used to assay the toxicity and uptake of NP-TEOS showing to be nontoxic without MTX making a good candidate for DDS.

INTRODUCTION

Recently, silica nanoparticles (NPs) have been intensively studied for their potential applications as drug delivery system [1, 2]. This system has several advantages including non-acute toxicity, biocompatible and biochemically inert. Also, its simplicity of preparation, by reaction from relatively inexpensive precursor molecules such as tetraethyl orthosilicate (TEOS) (Figure 1), methyl orthosilicate , polydimethylsiloxane and others [3],make theses nanoparticles unique with uniform mesopores which enable loading of large amounts of various chemical drugs [4]. The uniform small size allow easy uptake by lysosomes in cancer cells which size varies from 0.1 to 1.2 μm [5, 6]. Besides the size, another important aspect is its relation to pH dependence value between 4.8 and 6.4, substantially less than that of cytosol (pH = 7.4), which may be very favorable to drug release from a nanovalve system [2]. This type of DDS would ensure that the entrapped drug would be released from the nanoparticle only after the nanoparticle has been endocytosed into the target tissue [7].

MTX (Figure 1) is used to treat choriocarcinoma, leukemia in the spinal fluid, osteosarcoma, breast cancer, lung cancer, non-Hodgkin lymphoma, and head and neck cancers. It is also used to treat other cancers and non-cancerous conditions [8-10].This drug is also part of a general group of chemotherapy drugs known as *anti-metabolites*. It prevents cells from using folate to make DNA and RNA. Because cancer cells need these substances to make new cells, methotrexate helps to stop the growth of cancer cells. Despite its efficacy, the use of MTX is greatly limited due to its toxicity to normal cells, drug resistance, nephrotoxicity, bone marrow suppression, acute and chronic hepatotoxicity, interstitial pneumonitis and chronic interstitial obstructive pulmonary disease [10, 11]. To try to overcome these problems, different types of delivery vehicles for MTX were developed. In this study, we prepared MTX-incorporated in tetraethyl orthosilicate nanoparticles (NP-TEOS) to overcome these limitations.

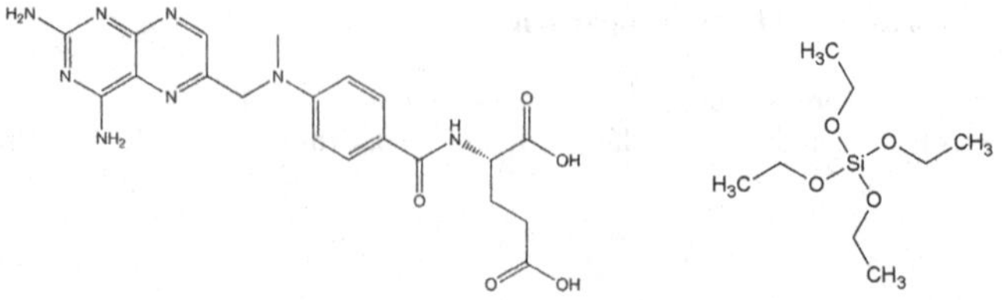

Figure 1. Chemical structure of Methotrexate (MTX) and tetraethyl orthosilicate (TEOS) respectively.

EXPERIMENT

MTX TEOS nanoparticles preparation

MTX was incorporated in silica nanoparticles using the optimized sol-gel process [12]. The first solution was a mixture of 200 μL TEOS and 16.0 mL methanol, and the second solution was a mixture of 4.0 mL 28 wt % ammonia, and 10.0 mg methotrexate (MTX). Under stirring (5,000 rpm), the first solution was added to the second solution, so that the total volume of the solution was about 20 mL. NPs were recovered by centrifugation at 15,000 rpm for 10 min at 4°C, washed with distilled water at 10°C, and lyophilized. Blank nanoparticles were prepared in a similar way so as to be used as a control in the characterization studies.

MTX TEOS nanoparticles characterization

In order to find the best condition of NP-TEOS suitable to further biological studies, various parameters were progressively modified to enhance the encapsulation of MTX in NP-TEOS. Once the appropriate size was reached, the physical chemical properties of the MTX-TEOS, such as the particle size distribution, zeta potential, drug entrapment efficiency and morphology were characterized. The particle size and morphology were assessed by photon correlation spectrometry (Zetasizer Nano ZS, Malvern Instrument, Worcestershire, UK) and scanning electron microscopy (SEM; ESEM 2020, Philips, Eindhoven, The Netherlands). The zeta potential (ζ) of the NPs (1.0 mg/mL) in 0.1 mM Hank's buffer (pH 7.4) was determined with a ZetaPlus (Malvern Instrument). The amount of MTX incorporated into the NP was estimated spectrophotometrically (U3900H, Hitachi, Tokyo, Japan) by means of its absorbance at 372 nm. The released MTX was quantified by the monitoring of its absorbance in the supernatant at 37° C. At determined intervals, the NP suspension was centrifuged at 15,000 rpm, for 10 min. For this, after centrifugation, the supernatant was evaporated at 25°C under nitrogen flux. After that, it was dissolved in an aliquot of 2.0 mL of sodium hydroxide (NaOH) 0.1 M, the absorbance spectrum was measured in the range between 250 and 600 nm. The amount of released material could be estimated with a standard curve. All measurements were done in triplicate.

The cell toxicity assay was performed in B16F10 cell line showing that E-NP-TEOS are nontoxic for cells in study.

RESULTS

The scanning electron microscope (SEM) is an excellent tool for observation of the external surfaces of nanoparticles. Figure 2 shows a representative micrograph of MTX loaded NP-TEOS magnified at 10,000x, were observed NP-TEOS with a mean diameter in the range of 270 at 400 nm. In all preparations reported in this study, the NP-TEOS were spherical in shape, displaying a smooth surface and show low tendency to aggregate. No meaningful difference was found between MTX load NP-TEOS and the empty nanoparticles (E-NP) used as a control, except for size and zeta variations.

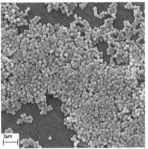

Figure 2. SEM pictures of MTX loaded NP-TEOS at magnification 10,000× (scale = 2 μm).

The characterization results of MTX loaded NP-TEOS showed an entrapment efficiency of 39.7 ± 4.0%, the size measurements (DLS) have a diameter between 140 and 430 nm (Figure 3) and between 149 and 237 nm in the empty NP-TEOS, with polydispersity index (PDI) of 0.12-0.26 indicated a narrow size distribution confirmed by SEM analysis.

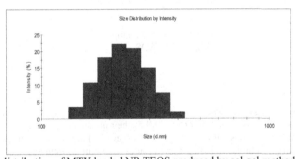

Figure 3. Size distribution of MTX loaded NP-TEOS produced by sol gel method

The colloidal stability was analyzed by measuring the NP-TEOS zeta potential values. The particles were negatively charged E-NP presenting -39.7 mV, whereas the zeta potential measured for the MTX loaded NP-TEOS was -61.4 ± 3.5 mV (Figure 4).

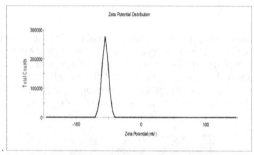

Figure 4. Zeta potential of MTX-TEOS produced by sol gel method

The release profile of the antitumoral MTX drug from nanoparticles showed that NP-TEOS possess a biphasic release pattern, which was characterized by an initial complex burst during the first 24 h, followed by a slower release phase complex profile (Figure 5), due to maybe a few pores or drug attached in the surface of NP-TEOS.

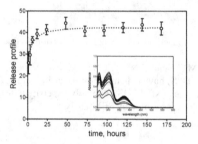

Figure 5. In vitro release profile of MTX from TEOS nanoparticles (mean ± SD; n = 3). Insert-Absorption spectrum of MTX released from NP-TEOS.

DISCUSSION

NP-TEOS is drug delivery system which could offer a number of advantages:(1) might help reduce undesirable side effects caused by drug; (2) they their physicochemical characteristics remain unaltered for long periods, allowing long-term storage; (3) depending on their composition they can be administered via different routes (oral, intramuscular, or subcutaneous); and (4) they are suitable for industrial production [13-15].

In this work SEM analysis (Figure 2) showed NP-TEOS with spherical shape, displaying a smooth surface, with a narrow size distribution (< 450 nm) and showed low tendency to aggregate. Also, the values obtained from DLS analysis support the SEM analysis. This behavior were also found in preparation of TEOS NP loaded with IFC-305, methylene blue, carvedilol [12, 16, 17].

According to the literature [18, 19] the administration of particles with a diameter of several micrometers seems to be inefficient as a DDS due to their accumulation in lung capillaries and the difficulty in removing them from the endothelial reticular system [19, 20]. Generally, nanoparticles have relatively higher intracellular uptake compared to microparticles [21]. The particle size significantly affects the level of cellular and tissue uptake, and in some cell lines, only submicron size particles are efficiently taken up [22]. According to the literature, other studies with particles (ranging from 10 to 4000 nm) loaded with different compounds, such as doxorubicin, loperamide, tubocurarine, and others have been carried out with the intent of injecting them intravenously [23]. In this respect, the particles size is important and should be characterized carefully.

Another important parameter evaluated is the zeta potential (ζ) of the surface of the NP, which indicates the tendency to aggregation. The particles were negatively charged, show ζ = -39.7 mV and -61.4 mV to E-NP and MTX loaded NP-TEOS respectively. The presence of MTX in NP-TEOS nanoparticles reduced the negative zeta potential value; probably, by deprotonation of hydroxyl groups on the silica particles and by the drug adsorbed on microparticles surface. This value is considered to be associated with a stable colloid [24]. In theory, more pronounced zeta-potential values, either positive or negative, tend to stabilize the particles suspension, since the electrostatic repulsion between particles with the same electric charge prevents aggregation of the spheres [20].

As shown in Figure 5, the biphasic pattern is observed for the drug release profile. This behavior can be attributed to the burst effect in the initial stage of the release, where the initial drug release phase is primarily controlled by the MTX diffusion, which takes place due to the dissolution of MTX molecules associated with the surface (*i.e.*, molecules that are not entrapped but are adsorbed on the surface of the NP). The burst release stage is followed by a slow release of MTX for up to 7 days, due to the low porosity of the NP-TEOS (Figure 2).

Our results showed that the MTX loaded NP-TEOS can be entrapped with preservation of its molecular structure and properties showed in the Uv-Vis analysis.

CONCLUSIONS

MTX loaded NP-TEOS prepared by modified method displayed important characteristics (size, zeta potential, release profile) suitable to be used as a potential drug delivery to MTX in cancer treatment.

ACKNOWLEDGMENTS

This work was supported by CNPq, CAPES, FAPDF and FINATEC

REFERENCES

1.   C. Barbe, J. Bartlett, L. G. Kong, K. Finnie, H. Q. Lin, M. Larkin, S. Calleja, A. Bush, and G. Calleja, *Advanced Materials*, **16**, 1959 (2004).
2. L. Du, H. Y. Song, and S. J. Liao, *Microporous and Mesoporous Materials*, **147**, 200 (2012).
3. A. Bitar, N. M. Ahmad, H. Fessi, and A. Elaissari, *Drug Discovery Today*, **17**, 1147 (2012).
4. J. E. Lee, D. J. Lee, N. Lee, B. H. Kim, S. H. Choi, and T. Hyeon, *Journal of Materials Chemistry*, **21**, 16869 (2011).
5. B. H. Kenzaoui, C. C. Bernasconi, S. Guney-Ayra, and L. Juillerat-Jeanneret, *Biochemical Journal*, **441**, 813 (2012).
6. W. Y. Zhai, C. L. He, L. Wu, Y. Zhou, H. R. Chen, J. Chang, and H. F. Zhang, *Journal of Biomedical Materials Research Part B-Applied Biomaterials*, **100B**, 1397 (2012).
7. P. DeMuth, M. Hurley, C. W. Wu, S. Galanie, M. R. Zachariah, and P. DeShong, *Microporous and Mesoporous Materials*, **141**, 128 (2011).
8. I. Sekine, H. Fukuda, H. Kunitoh, and N. Saijo, *Japanese Journal of Clinical Oncology*, **28**, 463 (1998).
9. K. P. Gao and X. G. Jiang, *International Journal of Pharmaceutics*, **310**, 213 (2006).
10.   D. H. Seo, Y. I. Jeong, D. G. Kim, M. J. Jang, M. K. Jang, and J. W. Nah, *Colloids and Surfaces B-Biointerfaces*, **69**, 157 (2009).
11.   M. A. Mashhadi, *Iranian Red Crescent Medical Journal*, **10**, 75 (2008).
12.   W. Tang, H. Xu, R. Kopelman, and M. A. Philbert, *Photochemistry and Photobiology*, **81**, 242 (2005).
13.   R. Diab, C. Jaafar-Maalej, H. Fessi, and P. Maincent, *Aaps Journal*, **14**, 688 (2012).
14.   S. Ivanov, S. Zhuravsky, G. Yukina, V. Tomson, D. Korolev, and M. Galagudza, *Materials*, **5**, 1873 (2012).
15.   L. Yildirimer, N. T. K. Thanh, M. Loizidou, and A. M. Seifalian, *Nano Today*, **6**, 585 (2011).
16.   L. Albarran, T. Lopez, P. Quintana, and V. Chagoya, *Colloids and Surfaces A-Physicochemical and Engineering Aspects*, **384**, 131 (2011).
17.   N. Andhariya, B. Chudasama, R. V. Mehta, and R. V. Upadhyay, *Journal of Nanoparticle Research*, **13**, 3619 (2011).
18.   A. J. Gomes, L. O. Lunardi, F. H. Caetano, A. E. H. Machado, A. M. F. Oliveira-Campos, L. M. Bendhack, and C. N. Lunardi, *Journal of Applied Polymer Science*, **121**, 1348 (2011).
19.   R. Kumar, I. Roy, T. Y. Ohulchanskky, L. A. Vathy, E. J. Bergey, M. Sajjad, and P. N. Prasad, *Acs Nano*, **4**, 699 (2010).
20.   A. J. Gomes, C. N. Lunardi, and A. C. Tedesco, *Photomedicine and Laser Surgery*, **25**, 428 (2007).
21.   A. J. Gomes, C. N. Lunardi, L. O. Lunardi, D. L. Pitol, and A. E. H. Machado, *Micron*, **39**, 40 (2008).
22.   M. R. Lorenz, V. Holzapfel, A. Musyanovych, K. Nothelfer, P. Walther, H. Frank, K. Landfester, H. Schrezenmeier, and V. Mailander, *Biomaterials*, **27**, 2820 (2006).
23.   A. D. Gomes, C. N. Lunardi, F. H. Caetano, L. O. Lunardi, and A. E. D. Machado, *Microscopy and Microanalysis*, **12**, 399 (2006).
24.   C. H. Lee, L. W. Lo, C. Y. Mou, and C. S. Yang, *Advanced Functional Materials*, **18**, 3283 (2008).

Mater. Res. Soc. Symp. Proc. Vol. 1498 © 2013 Materials Research Society
DOI: 10.1557/opl.2013.10

Bio-mimetic integrated surface nano structures for medical imaging scintillation materials

P. Pignalosa[1,2], B. Liu[3], W. Guo[2], X. Duan[4] and Y. Yi[2,4*]

[1]*New York University, New York, NY*
[2]*City University of New York, SI/GC, New York, NY*
[3]*Department of Physics, Tongji University, Shanghai*
[4]*Massauchusetts Institute of Technology, Cambridge, MA*

ABSTRACT
We have improved bio-inspired Moth eye nanostructures to enhance the scintillator materials external quantum efficiency significantly. As a proof of concept, we have demonstrated very high light output efficiency enhancement for Lu_2SiO_5:Ce^{3+} (LSO:Ce) film in large area. The X-ray mammographic instrument was employed to demonstrate the light output enhancement of the Lu_2SiO_5:Ce thin film with bio-inspired Moth eye-like nano photonic structures. Our work could be extended to other thin film scintillator materials and is promising to achieve lower patient dose, higher resolution image of human organs and even smaller scale medical imaging.
   *e-mail: yys@alum.mit.edu

INTRODUCTION
Inorganic scintillators are widely used in modern medical imaging modalities as converter for the x-rays and γ-radiation; they are also used to obtain information about the interior of the body, like x-ray, CT and PET scan [1-3]. Most of the high-density scintillators have a high refractive index, so when the light travels from inside the crystal to the air, the total reflection critical angle is small, most of light is trapped in the crystal, only 10-30% of the light from the scintillator can enter into the photodetector, the majority of light couldn't be effectively extracted, which seriously affected the detection system's efficiency and detection sensitivity. The use of micro and nano photonics can help us to explore new structural flashing material [4-8], such as artificial micro-nano structures, photonic crystals, optical microcavity, and surface plasmon materials. Artificial micro-nano structure can improve the extraction efficiency of light-emitting materials (LED) and is becoming more mature. However, the idea to apply this novel photonic structure for scintillator has just started [9-10].

In this work, we have investigated a novel class of nano scale devices based on photonic structures that function as efficient light extraction devices for scintillator materials. We have studied the light output enhancement characteristics of Lu_2SiO_5:Ce thin film scintillator materials, similar mechanism can be extended to various types of scintillator materials, such as γ-CuI single crystal and doped Tb^{3+} glass. We have designed light extraction structures based on their optical properties by using numerical simulation and optimized the sample preparation process, so that the overall efficiency of these scintillator materials is increased significantly.

With the increasing interest in reducing the front surface reflection, intensive effort has been focusing on the development of surface structures and process methodologies to achieve broadband antireflection [11-13]. Nano-structure comprising an array of circular protuberances (corneal nipples) on the facet lenses in Moth eye has been widely utilized to achieve this purpose. This bio-inspired nano structure is only one of many fascinating photonic structures in nature. Instead of broadband antireflection properties, this structure can also be utilized to enhance the light extraction of the scintillator materials.

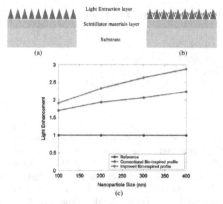

Fig. 1. The simulation device structure for two different bio-inspired photonic structures. (a) The conventional Moth eye structure, with flat sidewall. (b) The improved bioinspired Moth eye structure with roughness on the sidewall. (c) The FDTD simulation results, which shows light output enhancement for both bio-inspired light extraction structure, while the improved bio-inspired Moth eye structures shows more light output enhancement.

RESULTS

Due to our fabrication technique, we have further improved the bio inspired Moth eye structure for the light output enhancement, specifically, we have added some roughness on the sidewall of the original pyramid shape to increase the light scattering on the surface, and the comparison is illustrated in Fig. 1a and 1b. Fig. 1a is the conventional Moth eye structure with flat sidewall profile; Fig. 1b is the improved bio-inspired Moth eye structure, with certain degree roughness on the sidewall of the pyramid, with the average roughness around 50nm (similar to our fabricated structure in Fig. 2b). To demonstrate the advantage of the bio-inspired Moth-eye structure (both conventional and improved) for the scintillator materials, over the conventional structure, we have used Finite-Difference-Time-Domain (FDTD) method to study the light enhancement properties of the two structures. The device size is 100μm x 100μm, the grid size is 5nm and the Perfectly Matched Layer absorbing boundary condition (PBC) is used. Dipole sources with all polarizations (*x-dipole, y-dipole and z-dipole*) are used and averaged for the final simulation results. Fig. 1a and 1b are the device structure for Lu_2SiO_5:Ce thin film on glass substrate, the film thickness is 500nm, the high index light extraction material is Si_3N_4 with refractive index 2.0, and the height of the film is 300nm. The light enhancement results are shown in Fig. 1c, the blue curve is the reference structure with only Lu_2SiO_5:Ce thin film, the green curve is the light enhancement with conventional smooth Moth eye structure illustrated in Fig. 1a, and the red curve is the light enhancement with improved bio-inspired Moth eye structure with sidewall roughness illustrated in Fig. 1b. It can be clearly seen that the light enhancement factor from both light extraction structures is significant, while the improved bio-inspired Moth eye structure is larger than the original Moth eye structure, with the enhancement as large as 2.7 when the periodicity is 400nm. The external quantum efficiency is almost 86%, which demonstrates most of the light from the Lu_2SiO_5:Ce thin film has come out of the high index emission layer.

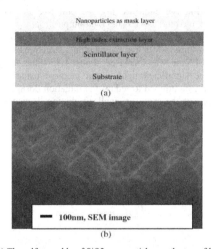

Fig. 2. (a) The self assembly of SiO2 nanoparticles on the top of high index light extraction layer Si3N4, which is deposited on Lu2SiO5:Ce thin film. (b) The SEM image of the improved bio-inspired moth-eye nanostructures with certain degree roughness on the sidewall, which shows interesting nano-on-nano features.

We have combined self assembly and reactive ion etching (RIE) methods to fabricate the improved bio-inspired Moth eye like nano photonic structures. The Lu_2SiO_5:Ce thin film, with nominal thickness 500nm, was deposited using sol-gel technique. As illustrated in Fig. 2a, the fabrication processes start from the coating of nominal 400nm SiO_2 nanoparticles, the SiO_2 nanoparticles were coated on the high index Si_3N_4 light extraction layer (whose refractive index is 2.0), which was previously deposited on the Lu_2SiO_5:Ce thin film. The nanoparticle layer on top of the Si_3N_4 film was used as a mask layer, the RIE method was then applied to etch the exposed Si_3N_4 film. After etching, the remaining SiO_2 nanoparticles were removed subsequently. Fig. 2b is the SEM image of the high index Si_3N_4 light extraction structure; it is very interesting to note the similarity to the Moth eye structure. In addition to the conventional smooth sidewall Moth eye structure, we notice the obvious *roughness* on the sidewall of each pyramid structure, and the *roughness* is at the same scale as the pyramid structure itself. Although our bio-mimetic structure is very similar to the Moth eye structures, the bumps (*roughness* at the similar scale) on the sidewall of the pyramid are showing new features, which will further enhance the light extraction properties. Furthermore, using self assembly of nanoparticles as mask is superior to conventional e-beam nanofabrication technique, for e-beam, it is very difficult to obtain large area nano scale patterns, with self assembly of

17

nanoparticles, we can potentially achieve reasonable large area nano scale patterns with real sample size as demonstrated in our experiment.

The X-ray mammographic unit was employed to demonstrate the light enhancement of the Lu_2SiO_5:Ce thin film with bio-inspired Moth eye-like nano photonic structures. The experiment was performed on a General Electric Senographe DMR plus X-ray mammographic unit with molybdenum anode target and molybdenum filter. Fig. 3 is our results on the comparison between the two Lu_2SiO_5:Ce thin films, the blue curve is the light output from the referenced Lu_2SiO_5:Ce thin film without any light extraction structures, the green curve is the light output from the Lu_2SiO_5:Ce thin film with improved bio-inspired Moth eye like photonic structures, as illustrated in Fig.2b. The peak light intensity at round 420nm was increased almost 1.75 times under the same dosage. Our results have demonstrated the significant light output increase using the improved bio-inspired photonic structures, which may be utilized to reduce the radiation dosage of the medical imaging or increase the image quality under the same or smaller radiation dosage for better and early stage diagnosis.

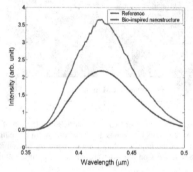

Fig. 3. The comparison between the two Lu2SiO5:Ce thin films; the curve below is the light output from the referenced Lu2SiO5:Ce thin film, and the curve above is the light output from the Lu2SiO5:Ce thin film with improved bio-inspired Moth eye like photonic structures.

SUMMARY

In summary, we have investigated a novel class of nano scale photonic structures that function as efficient light extraction devices for scintillator materials. The giant efficiency enhancement demonstrated using Lu_2SiO_5:Ce thin film can be extended to a family of scintillator materials. Our work shows the potential to study and understand a new class of nano photonic structures based on scintillator materials and devices.

REFERENCES

1. D. L. Bailey, J. S. Karp, S. Surti, Positron Emission Tomography, *Springer-Verlag London*, 29 (2005)

2. S. E. Derenzo, M. J. Weber, E. Bourret-Courchesne, M. K. Klintenberg, *Nuclear Instruments and Methods in Physics Research A*, **505**, 111(2003)

3. M. J. Weber, *Journal of Luminescence*, **100**, 35 (2002)

4. E. Yablonvitch, *Physical Review Letters*, **58**, 2059 (1987)

5. K. J. Vahala, *Nature*, **424**, 839 (2003)

6. J. Pendry, *Science*, **285**, 1687 (1999)

7. M. Boroditsky, T. F. Krauss, R. Coccioli, R. Vrijen, R. Bhat, E. Yablonovitch, *Applied Physics Letters*, **75**, 1036 (1999)

8. H. Ichikawa, T. Baba, *Applied Physics Letters*, **84**, 457 (2004)

9. M. Kronberger, E. Auffray, P. Lecoq, *IEEE Transactions on nuclear science*, **55**, 1102 (2008)

10. A. Knapitsch, E. Auffray, C. W. Fabjan, J.-L. Leclercq, P. Lecoq, X. Letartre, C. Seassal, *Nuclear Instruments and Methods in Physics Research A*, **628**, 385(2011)

11. C. H. Sun, P. Jiang, and B. Jiang, *Appl. Phys. Lett.*, **92**, 061112 (2008)

12. S. A. Boden, D. M. Bagnall, *Appl. Phys. Lett.* **93**, 133108 (2008)

13. H. Sai, Y. Kanamori, K. Arafune, Y. Ohshita, M. Yamaguchi, *Prog. Photovolt.: Res. Appl.* **15**, 415(2007)

Mater. Res. Soc. Symp. Proc. Vol. 1498 © 2013 Materials Research Society
DOI: 10.1557/opl.2013.56

Introducing Antibacterial Properties to Paper Towels Through the Use of Selenium
Nanoparticles

Qi Wang[1] and Thomas J. Webster[1,2]
[1]Bioengineering Program and [2]Department of Chemical Engineering, College of Engineering,
Northeastern University, Boston, MA 02115, U.S.A.

ABSTRACT

Bacterial infections are commonly found on paper towels and other paper products
leading to the potential spread of bacteria and consequent health concerns. The objective of this
in vitro study was to introduce antibacterial properties to paper towel surfaces by coating them
with selenium nanoparticles. Scanning electron microscopy was used to measure the size and
distribution of the selenium coatings on the paper towels. Atomic force microscopy was used to
measure the surface roughness of paper towels before and after coated with selenium
nanoparticles. The amount of selenium precipitated on the paper towels was measured by atomic
absorption spectroscopy. *In vitro* bacterial studies with *Staphylococcus aureus* were conducted to
assess the effectiveness of the selenium coating at inhibiting bacterial growth. Results showed
that the selenium nanoparticles coated on the paper towel surface were well distributed and
semispherical about 50nm in diameter. Most importantly, the selenium nanoparticle coated paper
towels inhibited *S. aureus* growth by 90% after 24 hours and 72 hours compared with the
uncoated paper towels. Thus, the study showed that nano-selenium coated paper towels may lead
to an increased eradication of bacteria to more effectively clean a wide-range of clinical
environments, thus, improving health.

INTRODUCTION

In the hospital environment, hand washing has been identified as the most significant
manner towards preventing the spread of microbial infections,[1,2] with hand drying as the critical
last stage of the hand washing process. Among the three frequently used methods to dry hands
(hot air dryers, cloth towels and paper towels), paper towels have been recognized as the most
hygienic method of hand drying.[3,4,5] However, in some circumstances, such as for paper towel
hanging in sink splash zones or those used to clean surfaces, they have been considered as
potential sources of bacteria contamination.[6]

Previously, studies evaluated the potential bacteria contamination of unused paper
towels.[7,8,9] In a hand wash experiment, participants who washed their hands with water, using
regular or antibacterial soap followed by drying with paper towels, surprisingly had more
bacteria on their hands after washing than before, which indicated a possible bacterial
transmission from paper towels.[8] It was further demonstrated that a zig-zag transfer of bacteria
between paper towel dispensers and hands could take place if either one is contaminated.[9]

Besides paper towels that are used for hand drying, there are concerns for many other paper products in terms of bacteria contamination or infection, for example food wrapping in the food industry[10], wall paper in a doctor's suite, filter paper in water purifying systems[11] and so on. All of those materials are prone to bacteria growth and, thus, are sources for continual contamination.

One of the most promising approaches towards preventing infections is coating paper products with antimicrobial materials. For example, Hu and his colleagues reported that introducing antibacterial properties to filter paper by coating the paper with graphene oxide, showed about a 70% inhibition to *Escherichia coli* growth after 2 hours. However, the graphene-based paper had mild cytotoxicity resulting in 20% of healthy mammalian A945 cell death after 2 hours.[12] Chule et al. studied the antibacterial activities of ZnO nanoparticle coated paper[13] and results showed a significant decrease in bacteria counts after 24 hours. But one major problem for ZnO and other metal-based materials is their toxicity to healthy cells due to the generation of reactive oxygen species.[15,16] Those materials may result in severe health problems when such coated paper products are used for food wrapping or clinical applications.

Compared with the above mentioned metal-based materials, selenium is considered to be more healthy and less toxic to healthy cells. In fact, it is recommended by the FDA that adults consume 53-60 μg of selenium, which contains the equivalent nutrition of 25 selenoproteins with selenocrysteine at their active center.[17] Therefore, in this study, for the first time, selenium nanoparticles were coated on normal paper towel surfaces through a quick precipitation method and the properties of selenium coated paper were characterized. In addition, their effectiveness to prevent biofilm formation was tested in bacterial assays involving *Staphylococcus aureus*. The results showed that the selenium coatings successfully introduced antibacterial properties to paper towels, revealing a promising selenium-based method to prevent bacterial infections on paper products that should be further explored.

MATERIALS AND METHODS

Materials and Materials Characterization

Paper towels (Tork Advanced, MB550A Hand Towel, cut into round chips, 7.01mm in diameter) were coated with selenium nanoparticles through a simple and quick precipitation reaction. The reaction involves glutathione (reduced form, GSH) (97%, TCI America, Portland, OR) and sodium selenite (99%, Alfa Aesar, Ward Hill, MA) mixed at a 4:1 molar ratio. Sodium hydroxide (0.5M) was added to bring the pH of the solution to the alkaline regimen, which favors the reaction. The coated substrates were rinsed in deionized water three times to remove the free, non-adherent, selenium nanoparticles and remaining reactants.

SEM (Scanning Electron Microscope, HITACHI 2700) images of the paper towel substrate surfaces were taken to determine the size, coverage and distribution of selenium nanoparticles. An AFM (Atomic Force Microscope, MFP3D, Asylum Research, sharp tipped cantilever, K = 0.06N/M, Contact Mode) was used to demonstrate that there was an increase in surface roughness on paper towel samples after being coated with selenium nanoparticles. The coated samples were treated in 1mL of aqua regia for 30 minutes to dissolve all the selenium into solution. After treatment, the solutions were collected in glass vials separately and then boiled to remove all the liquid. 5 mL of 2% nitric acid was added into each vial to dissolve the residue, which contained all the selenium from the solution. After about 24 hours, the solutions were measured with AAS (Atomic Absorption Spectroscopy, Furnace, AA600) to determine the concentration of selenium in each solution. Measurements were completed in triplicate for both blank control samples (uncoated paper towels) and selenium coated paper towel samples.

Bacterial Assays

A bacteria cell line of *S. aureus* was obtained in freeze-dried form from the American Type Culture Collection (catalog number 25923). The cells were propagated in 30 mg/mL tryptic soy broth (TSB) for 18 hours. A bacteria solution was prepared at a concentration of 10^6 bacteria/ml, which was assessed by measuring the optical density of the bacterial solution using a standard curve correlating optical densities and bacterial concentrations. The optical densities were measured at 562nm using a SpectraMax M5 plate reader (Molecular Devices, Sunnyvale, CA). The paper towel samples were transferred into a 24-well plate, treated with the prepared bacterial solutions (10^6 bacteria/ml), and cultured for either 24, 48 or 72 hours in an incubator (37°C, humidified, 5% CO_2). For those samples that were cultured for 48 and 72 hours, the media was changed with 1mL sterile and fresh TSB (0.3 mg/mL) every 24 hours. After the treatment, the samples were rinsed with a 10 mg/mL PBS (phosphate buffered saline) solution twice and placed into 1.5ml microfuge tubes with 1ml of PBS. These tubes were shaken at 3000 rpm for 15 minutes on a vortex mixer to release the bacteria attached on the surface into the solution. Solutions with bacteria were spread on agar plates and bacteria colonies were counted after 18 hours of incubation. Bacterial tests were conducted in triplicate and repeated three times. Bacterial tests were conducted in triplicate and repeated three times. Data were collected and the significant differences were assessed with the probability associated with a one-tailed Student's t-test. Statistical analyses were performed using Microsoft Excel (Redmond, WA).

RESULTS AND DISCUSSION

Paper Towel Characterization

Figure 1 shows the SEM images of the selenium coated paper towels (image a) and uncoated paper towels (image b). On the selenium coated paper towel samples, the selenium nanoparticles were well distributed and completely covered the surface. Some of the selenium nanoparticles were also observed in the fiber structure besides on the top surface. The diameters for most of the selenium particles were around 50 nm. For the uncoated paper towel, there were no particles observed. The AFM images showed that the RMS (root mean square, scan area = 10 $\mu m \times 10 \mu m$) roughness of the paper towel surface increased from 15.89nm (Figure 2b) to 31.14nm (Figure 2a) after coated with selenium nanoparticles. Thus, the selenium nanoparticles were successfully coated on the paper towels and the large surface area of the selenium coated fibrous paper towel surface increased the exposure of selenium. According to AAS results, the concentration of the selenium nanoparticles on the coated paper towel surface was 69.00 g/m^2..

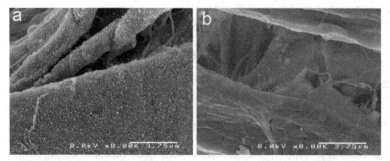

Figure 1. SEM images of selenium coated (image a) and uncoated paper (image b) towel samples. The coating condition for image (a) was 0.5M NaOH for 30 seconds. The concentration of selenium on the paper towel as measured by AAS was 69.00 g/m^2 for the selenium coated paper towels and 0 g/m^2 for the uncoated paper towels.

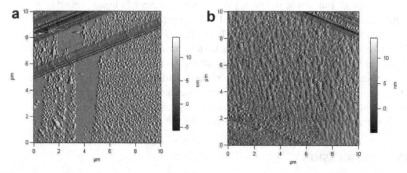

Figure 2. AFM images and RMS (root mean square, scan area = 10 μm×10 μm) roughness of paper towel surfaces. Image (a) is the surface of selenium coated paper towel with RMS roughness = 31.14 nm. Image (b) is the surface of uncoated paper towel with RMS roughness =15.89 nm.

<u>**Bacterial Assays**</u>

Most importantly, based on the bacterial assays, the selenium coated paper towel samples significantly inhibited biofilm formation compared with uncoated paper towel samples. As seen in Figure 3, the selenium coated paper towels had 88.6%, 88.9% and 88.8% less bacteria attached than the uncoated paper towels after 24, 48 and 72 hours, respectively. Moreover, from the 24 hour culture time to the 48 hour culture time, there was an increase in bacteria numbers on uncoated paper towel samples, but was constant to the 72 hour culture time, implying that the uncoated paper towel was saturated by bacteria after 48 hours of treatment. In contrast, the bacteria numbers on the selenium coated paper towels remained at a low level not increasing from 24 to 48 to 72 hours, indicating successful inhibition of bacterial growth. The surface with selenium nanoparticles revealed a stronger ability to prevent biofilm formation, especially when the bacteria in the biofilm propagating quickly. Overall, the bacteria growth and biofilm formation on paper towels were successfully inhibited after being coated with selenium nanoparticles.

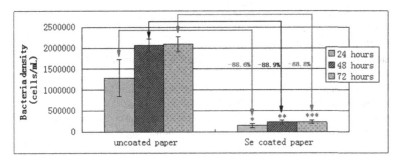

Figure 3. Bacteria (*S. aureus*) growth on the surface of uncoated and selenium coated paper towels. Data=Mean ± standard deviation, n=3; *p<0.02 compared with the control group (uncoated paper) after 24 hours; **p<0.002 compared with control group after 48 hours; ***p<0.002 compared with the control group after 72 hours.

CONCLUSIONS

In conclusion, the selenium precipitation process was used as an easy and quick method to coat selenium nanoparticles on paper towels, and the selenium coatings significantly inhibited the growth of *S. aureus* on the surface of paper towels after 24, 48 or 72 hours. The effectiveness of bacteria inhibition reached about 90% for all three different periods of treatment compared with the uncoated paper towels. This study, thus, suggests that selenium nanoparticles coatings could be used as an effective way to decrease *S. aureus* infections on paper products, which might have potentially important applications in the food packaging industry, medicine, and in clinical environments.

ACKNOWLEDGEMENTS

The authors thank Dr. Justin Seil for help with the bacteria experiment. They also thank Dr. Vera Fonseca for help with the AFM images.

REFERENCES

1. Garner JS. Favero MS. CDC guideline for handwashing and hospital environmental control, 1985. *ICHE*. 1986;7(4):231-243.
2. McGuckin M. Improving handwashing in hospitals: a patients education and empowerment program. *LDI Issure Brief.* 2001;7:1-4.
3. Guzewich J, Ross MP. White paper: evaluation of risks related to microbiological contamination of ready-to-eat food by food preparation workers and the effectiveness of interventions to minimise those risks. College park (MD): Food and Drug Administration, Centre for Food Safety and Applied Nutrition; September 1999.
4. Boyce JM, Pitter D. Draft guideline for hand hygiene in healthcare settings.*CDC/HICPAC*. 2002;51:No.RR-16.
5. Blackmore M . Hand drying methods. *Nurs Times*. 1987;83:71-4.
6. Hattula JL, Steven PE. A descriptive study of the handwashing environment in a long-term care facility. *Clin Nurs Res*. 1997;6:363-74.
7. Griffith CJ, Malik R, Cooper R, Looker N, Michaels B. Environmental surface cleanliness and the potential for contamination during hand washing. *Am. J. Infect. Control*. 2003;31:93-96.
8. Gendron LM, Trudel L, Moineau S, Duchaine C. Evaluation of bacterial contaminants found on unused paper towels and possible postcontamination after handwashing: A pilot study. *Am. J. Infect. Control*. 2012;40(2):E5-E9.
9. Harrison WA, Griffith CJ, Ayers T, Michaels B. Bacterial transfer and cross-contamination potential associated with paper-towel dispensing. *Am. J. .Infect. Control*. 2003;31(7):387-391.
10. Rodriguez A, Batlle R and Nerin C. The use of natural essential oils as antimicrobial solutions in paper packaging. Part II. *Progress In Organic Coatings*. 2007;60(1):33-38.
11. Yokota H, Tanabe K, et al. Arsenic contamination of ground and pond water and water purification system using pond water in Bangladesh. 2001;60(1-4):323-331.
12. Hu WB, Peng C, Luo WJ, et al. Graphene-based antibacterial paper. *ACS Nano*. 2010;4(7):4317-4323.
13. Chule K, Ghule AV, Chen BJ, Ling YC, et al. Preparation and characterization of ZnO nanoparticles coated paper and its antibacterial activity study. *Green Chemistry*. 2006;8(12):1034-1041.
14. Tankhiwale R, Bajpai SK. Graft copolymerization onto cellulose-based filter paper and its further development as silver nanoparticles loaded antibacterial food-packing material. *Colloids and Surfaces B: Biointerfaces*. 2009;69(2):164-168.
15. Xia T, Kovochich M, Nel AE, et al. Comparison of the mechanism of toxicity of zinc oxide and cerium oxide nanoparticles based on dissolution and oxidative stress properties. *ACS Nano*. 2008;2(10):2121-2134.
16. Foldbjerg R, Autrup H, et al. PVP-coated silver nanoparticles and silver ions induce reactive oxygen species, apoptosis and necrosis in THP-1 mono-cytes. *Toxicol Lett*. 2009;190:156-162.
17. Rayman MP. Selenium and human health. *The Lancet*. 2012;379(9822):1256-1268.

Mater. Res. Soc. Symp. Proc. Vol. 1498 © 2013 Materials Research Society
DOI: 10.1557/opl.2013.120

Effects of a Polycaprolactone (PCL) Tissue Scaffold in *Rattus norvegicus* on Blood Flow

Satish Bhat, [1] Christopher Chen, [2] Deborah A. Day[1]

[1] Science Research, Amity Regional High School, 25 Newton Road, Woodbridge, CT 06525
[2] Chemistry, University of Connecticut at Storrs, 97 North Eagleville Road, Storrs, CT 06269

ABSTRACT

Tissue engineering aims to save lives by producing synthetic organs and bone. This study is attempting to determine what effects a polycaprolactone (PCL) scaffold will have on the blood flow of Rattus norvegicus, as measured by the number of platelets. Prior to experimentation, it was hypothesized that the polycaprolactone scaffold would maintain and/or increase the number of platelets when compared to the control group. This was developed based on prior research that showed polylactic acid (PLA), a polymer being used currently, and polycaprolactone had similar characteristics like boiling point, melting point, and glass transition temperature. To test this hypothesis, the PCL, created from an existing protocol, was used to mold a scaffold in vitro. Three groups of rats were identified, then further split into an "A" and "B" subdivision with 5 members in each. All "A" subdivision members received the scaffold, while the "B" factions lacked it. Each rat underwent surgery to remove 1mm of the right ventricle, which was replaced by the PCL scaffold in the experimental group. The control group did not have the scaffold replacement. Without this piece of the right ventricle, prior research conducted at the University of Virginia in 2006 suggests that the rats would die within one week. However, in the experimental group of rats, the missing piece of the ventricle was replaced with the scaffold, so if it were accepted then the rats would survive beyond 1week. All rats in the experimental group died exactly 1 week after the control group as predicted before experimentation. After all of the rats had a 1-week acclimation period, a 1mm^2 slice of the heart was extracted and then the number of platelets was counted using a phase contrast microscope. The heart extraction was prepared in a petri dish and then placed into a hemocytometer, splitting the dish into smaller sections making it possible to count. The data supports the hypothesis whereby an average 12% increase in the number of platelets in the rats with the PCL scaffold versus the group without it was seen. This increase in platelet count reflects an increase in blood flow. A statistical t-test was conducted on each trial (n=5 per group, n=10 total per trial) comparing experimental versus control group to calculate a p-value. The p-values were 0.034, 0.045, and 0.022, respectively which indicates statistical significance since the value is less than 0.05. After all experimentation, the benefits of using PCL in tissue engineering were examined. For example, PCL costs $80 less to produce per kilogram than polylactic acid. This study suggests that PCL would be a viable candidate for tissue engineering in humans.

INTRODUCTION

Biomedical and tissue engineering are two budding fields of research. Biomedical engineering is a field in quest of narrowing the gap between medicine and engineering. The broad field of biomechanics and engineering is subcategorized into tissue engineering, genetic engineering, neural engineering, amongst others [1]. Commonly, tissue engineering is described as an interdisciplinary field which applies the principles of engineering toward the development of biological substitutes in order to restore, maintain, or improve tissue function [2]. These fields are key to the improvement of the organ donor system, which operates on a "first come-first served" basis [3]. Essentially, this means that people in need of organs may not survive because they were not "first in line" to receive an organ transplant [4]. Current research to develop artificial or synthetic organs such as this, may lend overall improvement in the system by creating organs ready for transplant in a more expedient and efficient way.

Among the major challenges facing tissue engineering is the need for more complex functionality, as well as functional and biomechanical stability. Specifically, tissue engineers are currently pioneering new ways to create organs via biological material for patients who need organ transplants. These specialists use techniques such as, nano-fiber self-assembly, textile technologies, and gas foaming to create scaffolds for implementation within the body. Currently, polylactic acid (PLA) is among the materials being used in tissue engineering. This study is designed to test a scaffolding material, polycaprolactone (PCL), to see whether or not it could be used in tissue engineering in place of PLA. To test this, a ring-opening polymerization of caprolactone to produce polycaprolactone (PCL) was done.

Polycaprolactone is a biodegradable polyester with a low melting point of approximately 60°C. PCL is a very cost-effective product, but it can be mixed with starch to lower its cost further and increase biodegradability [5]. However, in this study the polycaprolactone was not mixed with starch before implementing it as a scaffold. PCL is FDA (Food and Drug Administration) approved as a drug-delivery service and for sutures in the human body [6]. But, it is currently being investigated for use as a scaffold via tissue engineering as in this study [7].

It is understood that there is only a limited supply of polylactic acid. Therefore, the idea of having another material, polycaprolactone, which can be used, may be groundbreaking. This study is designed to investigate the effects of a polycaprolactone scaffold on the platelet count within *Rattus norvegicus*, as measured by the number of platelets in a 1mm x 1mm slice of the heart from the right ventricle calculated using a phase contrast microscope and hemocytometer.

Based on literature research, specifically a study done by the University of Michigan in 2005, it is hypothesized that polycaprolactone will be a viable material to be used for tissue engineering. The research in Michigan has shown that PCL scaffolds possess mechanical properties that may allow them to be used in functional loading [7]. Essentially this claim suggests that the use of a PCL scaffold as synthetic bone is promising. The researchers demonstrated that PCL had similar mechanical properties as bone, suggesting it as a viable material. In this specific experiment, further evidence to support the notion that polycaprolactone is a mechanically sound and viable material with possible applications in tissue engineering was sought.

EXPERIMENTAL DETAILS

2.1 Producing Polycaprolactone (PCL)

The polycaprolactone was produced as a result of a ring-opening polymerization of caprolactone. The Schwartz Catalyst (Cp_2ZrClH), initiator (terepthaldehyde), and solvent (toluene) were added to a test tube, which was then heated. After creating this mixture, the caprolactone was injected into the test tube and this new solution was spun, producing the polycaprolactone.

2.2 Creating Scaffold (Electrospinning)

The scaffold was created *in vitro* by using a technique known as electrospinning. Electrospinning uses an electric charge to extract very fine fibers from a liquid [9]. An SEM image was taken of the electrospun polycaprolactone fibers to ensure effectiveness.

2.3 Surgical Procedures

Each rat underwent a surgical procedure to remove a 1mm^2 piece of the right ventricle. Without this piece of the heart, the rats would not be able to survive beyond one week [10]. The experimental groups (A groups) had the piece of the heart that was extracted replaced with the PCL scaffold. The control group (B groups) rats did not receive this replacement. The scaffolds were allowed to acclimate in the rat bodies for a week before another 1mm^2 piece of the heart from the right ventricle was extracted to count the number of platelets.

2.4 Group Assignments

The 30 rats that were obtained for testing were split into three groups of 10 rats each, Groups 1, 2, and 3. There were 3 groups to measure if any difference could be seen between rats receiving and not receiving the scaffold. Also, it was to differentiate between the three trials that were conducted. These three major groups were then split further into A and B subgroups. The A subgroups would receive scaffold replacement, while the B subgroups would not.

2.5 Acceptance of Scaffolds

The scaffolds were tested for acceptance by a simple procedure. Prior research had stated that without the piece of the right ventricle that was removed, the rats would die naturally within 1 week [10]. If the rats with the PCL scaffold present survived beyond 1 week, the scaffold was determined as accepted and the rats were sacrificed by the researchers for evaluation.

2.6 Platelet Counts

The platelet counts were done using a hemocytometer and phase contrast microscope. The phase contrast microscope illuminated the platelets making it easier to count. The hemocytometer split the petri dish into smaller sections making it easier to count the platelets. The samples from each rat in both the A and B subgroups were counted and then averaged for a total of six measurements as seen in Table 1. The results were also graphed as seen in Figure 1.

Table I. Blood flow measurements as measured in method 2.6

	Presence of PCL	Acceptance of PCL	Platelets counted	Percent Increase/ Decrease
Group 1A	Yes	Yes	34	+30.78
Group 1B	No	-	26	-
Group 2A	Yes	Yes	42	+27.27
Group 2B	No	-	33	-
Group 3A	Yes	Yes	35	+9.38
Group 3B	No	-	32	-

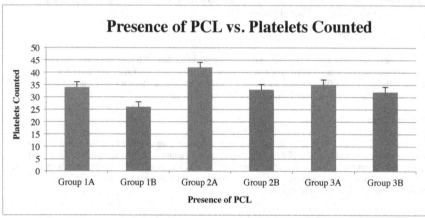

Figure 1. Blood flow measurements graphed as measured in method 2.6

III. DISCUSSION

In this study, the polycaprolactone showed promising results. The experimental groups, which had the PCL scaffold implemented in it continuously provided an increase in platelets through all three trials. For example, as seen in Figure 1 above, group one had a 30.78% increase in the number of platelets, from 26 to 34 over the course of the study. Furthermore, in group two there was a 27.27% increase in the number of platelets, from 33 to 42. In group 3, there was a 9.38% increase in the number of platelets, from 32 to 35. These differences in the average number of platelets among the three groups can be attributed to the difference in rats. It is near impossible to have the exact same average number of platelets in three completely different groups of 10 rats. These results show that the polycaprolactone may be a viable polymer for use in tissue engineering. The increase in platelet count throughout the three groups correlates to an increase in blood flow when the scaffold was present [5]. Thus, results from this study may offer an inexpensive material (PCL) to replace the current PLA in the tissue engineering field.

To support this, a cost analysis was conducted and showed that the PCL was 66% cheaper to produce than PLA. This would translate to saving $80 per kilogram produced of PCL versus PLA

IV. CONCLUSIONS

Prior to experimentation it was hypothesized that polycaprolactone would be a viable material to use in tissue engineering. This hypothesis was based on prior research that had investigated the idea of using polycaprolactone as scaffolding material used in bone regeneration [9]. After testing the implementation of PCL scaffolds within the heart of *Rattus norvegicus* this hypothesis is supported. The increase in platelets throughout all the trials when compared to the control group supports this.

ACKNOWLEDGEMENTS

The authors wish to thank Amity Regional High School and the University of Connecticut, Storrs for providing locations, resources, and opportunities to pursue this research.

REFERENCES:

1. Langer, R., Vacanti, J.P., "Tissue Engineering," Science 260:920-926 (1983).

2. MacArthur, B., Oreffo, R., "Concept Bridging the Gap," Nature 433:19 (2005).

3. Ritz, J., Fuchs, H., Kieczka, H., Moran, W., "Caprolactam," Encyclopedia of Industrial Chemistry (2011)

4. Venugopal, J., Zhang, Y.Z., Ramakrishna, S, "Fabrication of modified and functionalized polycaprolactone nanofibre scaffolds for vascular tissue engineering," Nanotechnology 16:2138 2005).

5. Wu, K.J., Wu, C.S., Chang, J.S., "Biodegradability and mechanical properties of polycaprolactone composites encapsulating phosphate-solubilizing bacterium *Bacillus* sp. PG01," Process Biochemistry 42:669-674 (2007).

6. Kweon, H., Yoo, M.K., Park, I.K., Kim, T.H., Lee, H.C., Lee, H.S., Oh, J.S., Akaike, T., Cho, C.S., "A novel degradable polycaprolactone networks for tissue engineering," Biomaterials 24:801-808 (2003).

7. Williams, J., Adewunmi, A., Schek, R., Flanagan, C., Krebsbach, P., Feinberg, S., Hollister, S., Das, S., "Bone tissue engineering using polycaprolactone scaffolds fabricated via selective laser sintering," Biomaterials 26:4817-4827 (2005).

8. Domingos, M., Dinucci, D., Cometa, S., Alderighi, M., Bártolo, P.J., Chiellini, F., "Polycaprolactone Scaffolds Fabricated via Bioextrusion for Tissue Engineering Applications," International Journal of Biomaterials 2009:1-10 (2009).

9. Lannutti, J., Reneker, D., Ma, T., Tomasko, D., Farson, D., "Electrospinning for Tissue Engineering Scaffolds," Materials Science and Engineering:C 27:504-509 (2007).

10. Sinha, V.R., Bansal, K., Kaushik, R., Kumria, R., Trehan, A., "Poly-ε-caprolactone microspheres and nanospheres: an overview," International Journal of Pharmaceutics 278:1-23 (2004).

Mater. Res. Soc. Symp. Proc. Vol. 1498 © 2013 Materials Research Society
DOI: 10.1557/opl.2013.159

Dextran Based Polyampholyte Having Cryoprotective Properties

Minkle Jain[1, 2] and Kazuaki Matsumura[1]

[1]School of Materials Science, Japan Advanced Institute of Science and Technology,

1-1 Asahidai, Nomi, Ishikawa 923-1292 Japan

[2]M. Tech (CSPT), Department of Chemistry, University of Delhi, Delhi-110007, India

ABSTRACT

Dimethyl sulfoxide (DMSO) has been used for several decades as the most efficient cryoprotective agent (CPA) for many types of cells in spite of its cytotoxicity and its effect on differentiation. Recently we showed that carboxylated poly-L-lysine, which is classified as a polyampholyte, has a cryoprotective effect on cells in solution without any other cryoprotectant. Here we developed high molecular weight polyampholytes with an appropriate ratio of amino and carboxyl groups and evaluated their cryopreservation efficiency. A novel polyampholyte based on naturally available polymer dextran, in which we introduced both amino and carboxyl groups shows an excellent post thaw-survival efficiency of more than 90% of murine L929 cells. It can serve as the sole high molecular weight CPA for tissue engineering applications without animal derived materials.

INTRODUCTION

The application of cryopreservation to living cells and tissues has revolutionized areas of biotechnology, plant and animal breeding programs, and modern medicine. It is the technique by which cells from prokaryotic to eukaryotic organisms can be recovered from temperatures down to almost two hundred degrees below the freezing point of water; this has been made possible by an important ingredient- the presence of cryoprotectant (CPA). CPA is the functionally-derived term coined to describe 'any additive which can be provided to cells before freezing and yields a higher post-thaw survival than can be obtained' in its 'absence' [1-3]. Dimethyl sulfoxide (DMSO), glycerol and other polyols have been used majorly for cryopreservation. However these current CPAs possess many problems. The cryoprotective properties of glycerol are relatively weak, and DMSO, although the most widely used CPA, shows high cytotoxicity [4] and affects the differentiation of neuron-like cells [5], cardiac myocytes [6] and granulocytes, and needs to be eliminated rapidly after thawing. Thus there is prerequisite need for the development of newer CPA. The success of tissue engineering applications in regenerative medicine requires advances in low-temperature preservation. However, cryopreservation of regenerated tissues including cell sheets and cell constructions is not easy compared to cell suspensions because of the weakness of cell-cell interaction and their inhomogeneous structures. Cryopreservation of cell-containing constructs is in high demand in tissue-engineering applications to produce the tissue engineered products "off-the-shelf". Recently we showed that carboxylated poly-L-lysine, which is classified as a polyampholyte, has a cryoprotective effect on cells in solution without any other cryoprotectant [7,8]. Cells are killed because of the damage caused by the intracellular crystallization of water during freezing. Therefore, a membrane-permeable chemical such as DMSO is usually added in order to cryopreserve the cells. However, the cryoprotective effect of polymers such as polyampholytes that do not penetrate the membrane cannot be explained by the same mechanism of DMSO. The recent study suggests that extra cellular environment might affect the cell viability after cryopreservation. We intend to develop a hydrogel forming CPAs as building blocks for cell scaffolds with cryoprotective properties for the application of tissue engineering.

In this study we attempted to make polyampholyte based on polysaccharide because of its facility of chemical modification to form hydrogels. We have shown that high molecular weight dextran based polyampholyte with an appropriate ratio of carboxyl and amino groups show high cryopreservation efficiency without the use of any animal derived proteins.

EXPERIMENTS

Amino-dextran preparation

To synthesize amino substituted dextran (NH$_2$-Dex), activation of hydroxyl groups of dextran (dextran, Mw 70000, Meito Sangyo Co., Ltd., Nagoya, Japan) was done by 1,1'-carbonyldiimidazole (Tokyo Chemical Industry CO., Ltd., Japan) for 2 h at 50 °C. In this reaction ethylenediamine (Wako Pure Chem. Ind. Ltd., Osaka Japan) was added after activation and the reaction was run for 20 h at 50 °C. NH$_2$-Dex was purified by dialysis (cutoff molecular weight: 14 kDa) against water for 48 h and then freeze-dried for 24 h. The amount of amino groups of NH$_2$-Dex was determined by the 2, 4, 6-trinitrobenzenesulfonate (TNBS) method [9] Briefly, 0.3 mL of 250 mg/mL sample solution, 1 mL of 1.0 mg/mL TNBS solution, and 2 mL of 40 mg/mL sodium bicarbonate aqueous solution containing 10 mg/mL sodium dodecyl sulfate (pH 9.0) were mixed and incubated at 37 °C for 2 h. After the mixture was cooled at 25 °C, the absorbance was measured at 335 nm.

Polyampholyte Preparation

To synthesize carboxylated amino-dextran (dextran polyampholyte, Dex-PA), NH$_2$-Dex and succinic anhydride (SA) (Wako) in 0–90% mol ratios (SA/NH$_2$-dextran amino groups) were mixed and reacted at 50 °C for 3 h to convert amino groups into carboxyl groups.

Cell culture

L929 cells (American Type Culture Collection, Manassas, VA, USA) were cultured in Dulbecco's modified Eagle's medium (DMEM, Sigma–Aldrich, St. Louis, MO) supplemented with 10% fetal bovine serum (FBS). Cell culture was carried out at 37 °C under 5% CO$_2$ in a humidified atmosphere. When the cells reached 80% confluence, they were removed by 0.25% (w/v) of trypsin containing 0.02% (w/v) of ethylenediamine tetra acetic acid (EDTA) in phosphate buffered saline without calcium and magnesium (PBS (-)) and seeded on a new tissue culture plate for subculture.

Cryopreservation

Dex-PA cryopreservation solutions were prepared as follows. Dex-PA dissolved in DMEM at the concentration of 7.5-15% (w/w) and pH was adjusted to 7.4 using HCl or NaOH. Osmotic pressure was adjusted by addition of 10 w/w% of NaCl aqueous solution to about 550 mOsm using an Osmometer (Osmometer 5520, Wescor, Inc., UT, USA). L929 cells were counted and resuspended in 1 mL of polyampholyte solution at a density of 1×10^6 cells/mL in 1.9 mL cryovials and stored in an -80 °C freezer overnight. Individual vials were thawed at 37°C in a water bath with gentle shaking, and the thawed cells were diluted 10-fold in 4 °C DMEM. After centrifugation, the supernatant was removed and the cells were resuspended in a 5 mL medium. All the cells were counted using a hemocytometer with a trypan blue staining method. The reported values are the ratios of living cells to total cell numbers.

DISCUSSION

Amino groups were introduced into dextran by treatment with ethylenediamine (figure 1a), and degree of substitution (DS) per sugar unit was 34.8% calculated by TNBS assay. The amino groups were converted into carboxyl groups by treatment with SA (figure 1b). The ratio of carboxylation shown in parentheses, e.g., Dex-PA (0.70), indicates that 70% of the amino groups introduced into dextran have been converted into carboxyl groups by SA addition. Figure 2a shows cell viabilities immediately after thawing with 10% polymer concentration solution of Dex-PA with various COOH introduced ratios. Cell viability increased with the percentage of introduced carboxyl groups up to 70%. When L929 cells were preserved with 10% Dex-PA (at carboxylation ratios of 0.50, 0.60, 0.65, 0.70, 0.75, 0.8 and 0.9) viability after thawing was similar to that of 10% DMSO without FBS. When Dex-PA (0.70) was used for cryopreservation, concentrations of 10% resulted in highest viability(figure 2b). We controlled the osmotic pressure of the Dex-PA at about 560mOsm because of the results of a viability assay against the osmotic pressure of Dex-PA (0.70). In this assay, Dex-PA (0.70) solutions with relatively higher osmotic pressures of 400–700 mOsm compared to that of the living body exhibited better cryopreservation properties (figure 3) than those with lower pressures. It is likely that the cells were dehydrated under relatively high osmotic pressure during cooling, and thus avoided damage from intracellular freezing [10-12]. However, an excessive high osmotic pressure (1403 mOsm) heavily damaged cells. More than 90% viability was obtained in cryopreservation with 10% Dex-PA (0.70) with 560mOsm.

These results showed that dextran ethylenediamine based polyampholytes have cell cryoprotective property similar to poly-L-Lysine [7,8] and that cryoprotection may be a property of not only polylysine derivatives but also polyampholytes generally.

Also from the result that such a high molecular weight (70000) polyampholytes could cryopreserve cells well, we will develop strategy of hydrogel formation of these polyampholytes by chemical modification for cell scaffolds having cryoprotective properties in the future study.

a

Amino-Dex

b

COOH-Dex-NH₂
Polyamholyte

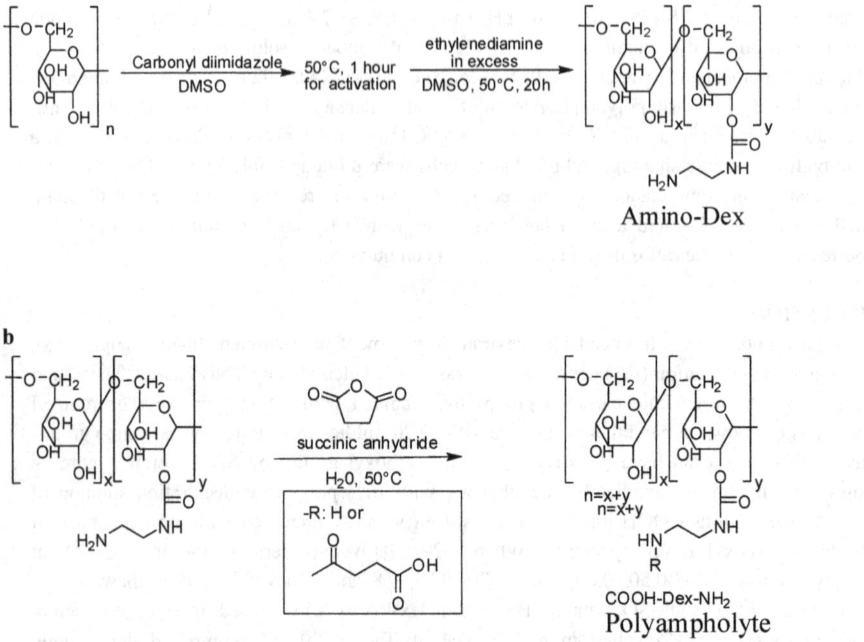

Fig.1 Synthesis of Dex-PA. (a) Schematic representation of reaction of amination of dextran. (b) Schematic representation of reaction of succinylation of NH_2-dextran.

a

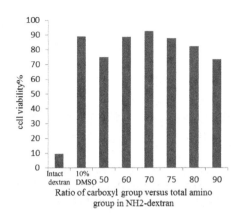

b

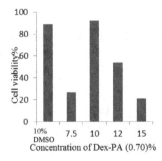

Fig.2. Cryoprotective properties of Dex-PAs. (a) L929 cells were cryopreserved with 10% DMSO and 10% (w/w) Dex-PAs with different ratios of introduced COOH. (b) L929 cells were cryopreserved with various concentrations of Dex-PA (0.70).

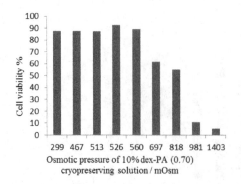

Fig.3 The viability after thawing of L929 cryopreserved with Dex-PA (0.70) as a CPA. L929 cells were cryopreserved with 10% Dex-PA (0.70) under various osmotic pressures. Osmotic pressures were adjusted with 10% (w/w) NaCl aqueous solution.

CONCLUSIONS

Dextran based polyampholytes show post-thaw survival efficiency of more than 90% of L929 murine cells. Thus, they can replace conventional used CPAs. Further investigation has to be done in the future to provide the hydrogel forming CPAs by chemical modification for tissue engineering application.

REFERENCES

1. J.O.M. Karlsson, M. Toner, *Biomaterials*, **17**, 243–256 (1996).
2. A.M. Karow Jr. and W.R. Webb, *Cryobiology* **1,** 270-273(1965).
3. A.M. Karow Jr., *J Pharm Pharmacol*, **21**, 209-223(1969).
4. G.M. Fahy, *Cryobiology*, **23**, 1–13 (1986).
5. J.E. Oh, K. Karlmark Raja, J.H. Shin, A. Pollak, M. Hengstschlager, G. Lubec, *Amino Acids*, **31**, 289–98 (2006).
6. D.A. Young, S. Gavrilov, C.J. Pennington, R.K. Nuttall, D.R. Edwards, R.N. Kitsis et al., *Biochem. Biophys. Res. Commun.*, **322**, 759–65 (2004).
7. K. Matsumura, S.H. Hyon, *Biomaterials* **30**, 4842-4849 (2009).
8. K. Matsumura, J.Y. Bae, S.H. Hyon, *Cell Transplant.* **19**, 691-699 (2010).
9. A.F. Habeeb, *Anal Biochem*, **14**, 328–36 (1966).
10. P. Mazur, *Cryobiology*, **14**, 251–72 (1977).
11. P. Mazur, I.L. Pinn, F.W. Kleinhans, *Cryobiology*, **5**, 158–66 (2007).
12. M. Toner, E.G. Cravalho, J. Stachecki, T. Fitzgerald, R.G. Tompkins, M.L. Yarmush et al., *Biophys J*, **64**, 1908–21 (1993).

Mater. Res. Soc. Symp. Proc. Vol. 1498 © 2012 Materials Research Society
DOI: 10.1557/opl.2012.1678

Effect of Substrate Elasticity on *In Vitro* Aging of Human Mesenchymal Stem Cells

Courtney E. LeBlon,[1] Caitlin R. Fodor,[2] Tony Zhang,[2] Xiaohui Zhang,[1] Sabrina S. Jedlicka[2,3,4]

[1]Mechanical Engineering & Mechanics, [2]Bioengineering Program, [3]Materials Science & Engineering, [4]Center for Advanced Materials & Nanotechnology
Lehigh University, Bethlehem, PA

ABSTRACT

Human mesenchymal stem cells (hMSCs) were routinely cultured on tissue-culture polystyrene (TCPS) to investigate the *in vitro* aging and cell stiffening. hMSCs were also cultured on thermoplastic polyurethane (TPU), which is a biocompatible polymer with an elastic modulus of approximately 12.9MPa, to investigate the impact of substrate elastic modulus on cell stiffening and differentiation potential. Cells were passaged over several generations on each material. At each passage, cells were subjected to osteogenic and myogenic differentiation. Local cell elastic modulus was measured at every passage using atomic force microscopy (AFM) indentation. Gene and protein expression was examined using qRT-PCR and immunofluorescent staining, respectively, for osteogenic and myogenic markers. Results show that the success of myogenic differentiation is highly reliant on the elastic modulus of the undifferentiated cells. The success of osteogenic differentiations is most likely somewhat dependent on the cell elastic modulus, as differentiations were more successful in earlier passages, when cells were softer.

INTRODUCTION

hMSCs are adult stem cells that can differentiate into a variety of lineages, including bone, cartilage, fat, tendon, muscle, and marrow stroma.[1] hMSCs are able to proliferate in culture while retaining their multilineage potential. Because of their multipotency, hMSCs have great potential for regenerative medicine and tissue engineering. However, MSCs only represent 0.01% to 0.001% of cells in bone marrow.[2] Therefore, to produce sufficient cell numbers for therapeutic use, hMSCs must be expanded *in vitro*.

TCPS is the standard growth substrate in cell culture laboratories. It has an elastic modulus of approximately 3 GPa, is several orders of magnitude stiffer than most values for cells, which are typically in the kPA range.[3-6] This stiff culture substrate is likely not the ideal material for stem cell maintenance.

hMSCs are subject to *in vitro* aging, which is hallmarked by telomere shortening, slowed proliferation, and decreased differentiation capacity.[5, 7-11] For example, Bonab et al. studied the osteogenic and adipogenic differentiation potential of expanded hMSCs. By the 10th passage *in vitro*, 20% and 60% of the samples lost their osteogenic and adipogenic differentiation potential, respectively.[7] Aging *in vitro* is accompanied by changes in the mechanical properties of hMSCs, most likely due to extremely stiff *in vitro* culture substrates, as it has been shown that the elasticity of hMSCs can be modulated by substrate.[12]

To examine the mechanical properties of cells, AFM indentation is commonly used.[13-15] Maloney et al. used AFM to indent hMSCs maintained on TCPS over 17 population doublings and saw the stiffness increase from approximately 2 kPa to 8 kPa.[5] Correlating to differentiation capacity, the cells also exhibited a reduction in differentiation potential for osteogenic and adipogenic differentiation between population doublings 4 and 14. Given these observations and others, it seems evident that cell culture conditions are contributing to the *in vitro* aging effects of cell stiffening and subsequent decreased differentiation capacity.

A softer growth substrate may be more suitable for long-term *in vitro* culture to prevent cell aging. Because hMSCs have been shown to differentiate on gels with elastic modulus values in the kPa range (in the absence of inductive chemicals),[16] a material in the MPa range was investigated. A TPU, TexinRxT85A was injection molded to create thin (1mm) membranes. The tensile modulus at room temperature (ASTM D635) is 12.9MPa. It does not biodegrade, so it is suitable for long-term stem cell maintenance. Since it can be injection molded, cell culture substrates can be easily reproduced.

This study examines how differentiation capacity changes with elastic modulus of undifferentiated hMSCs that have been cultured on both TCPS and TPU. AFM was used to quantify cell elastic modulus over many *in vitro* passages. At various passages, osteogenic and myogenic differentiations were carried out. Immunocytochemistry and qRT-PCR were used to assess the success of the differentiations.

EXPERIMENTAL DETAILS

Cell culture

hMSCs were cultivated on TCPS dishes and TPU membranes in Mesenchymal Stem Cell Basal Medium supplemented with MSCGM SingleQuots (Lonza, Walkersville, MD). When cells reached 70% confluency, they were split 1:6 (approximately 1000 cells/cm^2). To induce osteogenic differentiation, hMSCs were incubated in osteoinductive media for 4 weeks. Osteoinductive media contained DMEM Low Glucose with L-glutamine, glucose, and sodium pyruvate supplemented with 10% fetal bovine serum (FBS), 0.1μM dexamethasone, 10 mM β glycerol phosphate, and 0.05 mM ascorbic acid. To induce myogenic differentiation, 5 mM azacytidine was added to media (DMEM Low Glucose, 10% FBS) once a week for 24 hours. This procedure was repeated for 4 weeks.

AFM indentation

Local cell elastic modulus was measured at every passage using atomic force microscopy (AFM) indentation. A Basal Locke Buffer was used during indentation to maintain pH. Fifty cells were indented per passage, and cells were indented in 2 spots (1 on the edge of the cell and 1 in the cytoplasm). Silicon nitride cantilevers with spring constant of 0.01 N/m were used and calibrated before each experiment. A maximum signal of 500 mV resulted in an indentation depth of approximately 250 nm. The sample rate was 8000 Hz. An effective elastic modulus of the indented cell is estimated by applying the Hertz equation (Equation 1), where F is the loading force, δ is the indentation depth, E is the elastic modulus, ν is the Poisson ratio (assumed to be 0.5 for cells), and α is the opening angle of the cone tip (15°).[17]

$$F = \delta^2 \frac{2}{\pi} \frac{E}{(1-v^2)} \tan\alpha \qquad (1)$$

Data was analyzed custom-made program written in IgorPro (WaveMetrics, Portland, OR). Statistical significance was evaluated with a paired Student's t-test. Significance levels were set at *P* values of 0.1, 0.05, and 0.01. All values are reported as means ±SE unless otherwise stated.

Actin stress fiber measurements

The average stress fiber diameter was calculated at various population doublings using fluorescence microscopy. hMSCs were fixed in 10% formalin and stained with rhodamine phalloidin to view actin stress fibers. The images were analyzed using ImageJ. Fifty cells were analyzed for every passage. The experiment was repeated.

Immunocytochemistry

Fixation and immunocytochemical processing for osteogenic and myogenic markers was conducted using standard protocols. The monoclonal antibodies mouse anti-osteopontin (OP) (MPIIIB10(1), 1:500), developed by Michael Solursh and Ahnders Franzen, and anti-osteonectin (ON) (AON-1, 1:500), developed by John D. Termine, were obtained from the Developmental Studies Hybridoma Bank developed under the auspices of the NICHD and maintained by The University of Iowa, Department of Biology, Iowa City, IA 52242. Phycoerythrin-conjugated mouse anti-osteocalcin (OC) (1:500) was purchased from R&D (Minneapolis, MN). Sheep polyclonal anti-tropomyosin (1:500) was purchased from Millipore (Billerica, MA). Mouse monoclonal anti-sarcomeric actin (1:250) and FITC-conjugated CD44 (1:100) were purchased from Invitrogen (Carlsbad, CA). Briefly, cells were fixed with 10% formalin in 1 x PBS for 15 minutes. The samples were then treated with 100% methanol for 7 minutes. Non-specific binding sites were blocked for 30 minutes with agitation at room temperature with 1% bovine serum albumin (BSA). Primary antibodies were incubated for 1h at 37°C, followed by incubation with appropriate secondary antibodies (Alexa Fluor 488, 546, or 555, 1:1000, Invitrogen) for 1h at room temperature. The nuclei were then counterstained with Hoechst dye (0.002 mg/ml in 1xPBS) for 5 minutes.

qRT-PCR

qRT-PCR was performed at various population doublings to assess RNA expression. Total RNA was isolated using Qiagen Mini-Prep kits. Cycles were optimized for the amplicon size and primer Tm on a Qiagen Rotor-Gene qRT-PCR system. A SYBR green qRT-PCR kit from Qiagen was used to assess gene expression. The following genes were analyzed: GAPDH, smooth muscle α-actin (SMAA), calponin 1 (CNN1), desmin, β-myosin heavy chain (MHC), Runx2, OC, Collagen I (Col I), alkaline phosphatase (ALP), and ON. Primers available upon request.

Analysis of the qRT-PCR data was completed using the $\Delta\Delta$Ct method, with a threshold value of 20% above the background. Efficiency values were empirically derived from undifferentiated hMSCs for the housekeeping gene (GAPDH), osteogenic hMSCs for bone markers, and myogenic hMSCs for muscle markers. Fold changes were determined through $\Delta\Delta$Ct comparison to the negative control. All data was subject to normalization to the housekeeping gene (GAPDH) corresponding to each experimental condition and specimen.

RESULTS

AFM indentation and actin stress fiber measurements

Prior to differentiation, the elastic modulus of hMSCs on TCPS was analyzed (Figure 1a). The elastic modulus of undifferentiated hMSCs remained fairly constant from P3 to P5 (approximately 8kPa). Significant increases ($P<0.05$) were observed from P6 (9.65 ± 1.13 kPa) to P7 (12.97 ± 1.26 kPa), and from P7 to P8 (17.42 ± 1.46 kPa). After P8, the average elastic modulus value was 19kPa. These results are consistent with other studies, with any differences due to donor age, AFM tip geometry[18] and location of indentation, as elastic modulus values vary widely in a single cell.[19] The actin stress fiber diameters of undifferentiated cells on TCPS can be seen in Figure 1b. The diameter increased with every passage, which led to increases in cell elastic modulus. Very significant (P<0.01) increases occurred between each passage from P5 to P9.

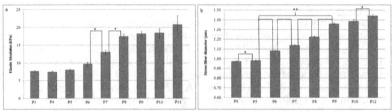

Figure 1. a) Change in elastic modulus from P3 to P11, b) Change in actin stress fiber diameters from P4 to P12. (*P<0.05, **P<0.01)

Myogenic differentiations

Myogenic differentiations were carried out at every passage. The protein expression of two myogenic markers, tropomyosin and sarcomeric actin, were examined using immunocytochemistry (Figure 2). Tropomyosin is a protein that regulates sarcomeric contraction. Tropomyosin interacts with actin and the troponin complex to control the attachment of crossbridges to actin.[20] Sarcomeric actin begins to appear upon the appearance contraction in cardiac and skeletal muscle cells.[21, 22] Protein levels of tropomyosin and sarcomeric actin peaked at P7, but were barely visible prior to P6 and after P8. At the passage of peak expression, P7, the elastic modulus value was 12.97±1.26 kPa. It appears that this value is most conducive to myogenic differentiation of hMSCs. Engler et al. previously reported that the elasticity of skeletal muscle is approximately 12kPa.[27]

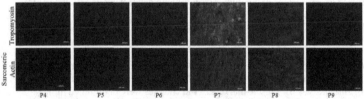

Figure 2. Expression of myogenic markers in differentiated hMSCs from P4 to P9.

mRNA expression in differentiated hMSCs on TCPS was investigated using qRT-PCR (Table I). mRNA levels of desmin and SMAA peaked at P6. Desmin is a muscle-specific intermediate filament that appears in the early formation of cardiac, skeletal, and smooth muscle. Desmin is thought to be involved in myofibrillogenesis and mechanical support for muscle cells.[23] SMAA, the primary actin isotype found in vascular smooth muscle cells, is also transiently expressed in the development of cardiac and skeletal muscle.[24] Levels of CNN1 and MHC peaked at P7 and P8, respectively. CNN1 is a protein that interacts with F-actin and tropomyosin to control the contraction of smooth muscle cells.[25] Myosin is the main component of thick filaments in sarcomeres. Type II myosin, found in muscle cells, has two heavy chains and four light chains. The heavy chains are each comprised of a globular head domain and a coiled tail domain.[26] In this study, a cardiac muscle-specific MHC isoform was used. It is interesting to note that the cells express markers associated with all types of muscle. However, CNN1, a marker associated with smooth muscle, was much more upregulated than the cardiac muscle-specific marker MHC. The increased upregulation of myogenic markers from P4 to P7, peak protein expression of sarcomeric actin and tropomyosin at P7, and the stiffening of cells to an optimal elasticity (approx. 13kPa) all seem to contribute to the success of myogenic differentiation.

Table I. qRT-PCR results for myogenic hMSCs.

Passage	SMAA	CNN1	MHC	Desmin
4	8.05	4.43	0.03	137.21
5	10.47	14.61	0.02	550.83
6	35.56	22.38	0.01	2257.04
7	33.43	178.64	1.17	1046.64
8	21.20	12.83	8.00	-
9	17.72	3.20	0.21	-

Osteogenic differentiations

Osteogenic differentiations were carried out at every passage. ON, OP, and OC expression were assessed with immunocytochemistry (Figure 3). ON is an extracellular matrix glycoprotein that initiates mineralization through binding to collagen and hydroxyapatite.[28] OP is a secreted protein that is essential for some forms of bone remodeling.[29] OC is involved in regulation of mineral deposition and is a marker of mature osteoblasts.[30] ON protein expression appears to decrease after P4. OP and OC protein expression were minimal by P7 and P8.

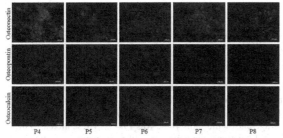

Figure 3. Expression of osteogenic markers in differentiated hMSCs from P4 to P8.

qRT-PCR results for osteogenic hMSCs on TCPS can be seen in Table II. Large decreases in Runx2 and ALP were seen after P4, although both genes remained upregulated through P8. Runx2 is an essential transcription factor in osteogenesis that regulates OC and OP expression.[31] ALP is a metalloenzyme that initiates mineralization, and peaks early in differentiation.[32] Col 1 is the main collagen protein in the bone matrix, and its gene expression typically peaks in the first 2 weeks of differentiation.[33] Col 1 was downregulated at P6. OC and ON gene expression were downregulated at P7 and P8. These results suggest that osteogenic differentiation may be incomplete for P7 and P8, given OP and OC protein expression were also minimal.

It is unclear whether the success of osteogenic differentiations is reliant on the elastic modulus of the undifferentiated cells. However, there are large increases in elastic modulus from P6 (9.65±1.13 kPa) to P7 (12.97±1.26 kPa) and P8 (17.42±1.46 kPa), which are the passages where several key players in osteogenesis are downregulated. Additionally, Titushkin et al. demonstrated that the elastic modulus actually decreased when hMSCs became fully differentiated osteoblasts (3.2±1.4 kPa to 1.7±1.0 kPa).[6] This suggests that hMSCs may have more differentiation success in earlier passages, when they exhibit reduced endogenous cell stiffness.

Table II. qRT-PCR results for osteogenic hMSCs.

Passage	Runx2	OC	Col I	ALP	ON
4	1152.38	9.34	3.34	1546.67	2.57
5	165.34	18.00	7.50	717.34	3.70
6	282.67	2.33	0.21	518.00	4.01
7	20.00	0.88	0.073	512.00	0.75
8	64.00	0.79	0.15	469.33	0.0041

hMSCs on TPU

TPU is nonbiodegradable, biocompatible polymer that was injection molded to create thin (1mm) membranes that fit into cell culture dishes. Adherent hMSCs can be seen in Figure 4. Since hMSCs adhered to this substrate, it has potential to be a long-term culture material. Because of its lower modulus (12.9MPa), it may prevent cell stiffening.

Figure 4. hMSCs on injection molded TPU membrane.

CONCLUSIONS

Cells passaged on TCPS were subject to cell stiffening. The elastic modulus of undifferentiated cells remained constant (approximately 8kPa) from P3 to P5. Significant increases ($P<0.05$) were observed from P6 (9.65±1.13 kPa) to P7 (12.97±1.26 kPa) and P8 (17.42±1.46 kPa). After P8, the average elastic modulus value was 19kPa. These increases in cell elastic modulus were accompanied by increases in actin stress fiber diameters. Elastic modulus values of hMSCs on TCPS largely dictated the success of myogenic differentiation. Highest expression of myogenic markers were observed at P7, where E=12.97±1.26 kPa, which is comparable to previously reported values for muscle cells. Osteogenic differentiations on TCPS were more successful in earlier passages (P4 through P6), which may suggest that hMSCs prefer a softer state. Cells adhered to injection molded TPU membranes, which makes them a candidate for long-term cell culture. This softer cell substrate may prevent the cell stiffening and aging that occurs, and studies to this effect are currently underway.

ACKNOWLEDGEMENTS

The research was supported by NSF Grant #1014987. We also acknowledge the Lehigh University Faculty Innovation Grant (2009-2010).

REFERENCES
1 M. F. Pittenger *et al.*, Science 284 (5411), 143-147 (1999).
2 AJ Friedenstein *et al.*, Exp.Hematol. 10 (2), 217-227 (1982).
3 M. Radmacher *et al.*, Biophys.J. 70 (1), 556-567 (1996).
4 Eric M. Darling *et al.*, J.Biomech. 41 (2), 454-464 (2008).
5 John M. Maloney *et al.*, Biophys.J. 99 (8), 2479-2487 (2010).
6 Igor Titushkin and Michael Cho, Biophys.J. 93 (10), 3693-3702 (2007).
7 MM Bonab *et al.*, BMC Cell Biol. 7, 14 (2006).
8 A. Banfi *et al.*, Exp.Hematol. 28 (6), 707-715 (2000).
9 S. P. Bruder, N. Jaiswal and S. E. Haynesworth, J.Cell.Biochem. 64 (2), 278-294 (1997).
10 Shuanhu Zhou *et al.*, Aging Cell 7 (3), 335-343 (2008).
11 Jiseon Kim *et al.*, Arch.Pharm.Res. 32 (1), 117-126 (2009).
12 Evelyn K. F. Yim *et al.*, Biomaterials 31 (6), 1299-1306 (2010).
13 U. G. Hofmann *et al.*, J.Struct.Biol. 119 (2), 84-91 (1997).
14 A. B. Mathur *et al.*, J.Biomech. 34 (12), 1545-1553 (2001).
15 M. Radmacher, IEEE Eng.Med.Biol.Mag. 16 (2), 47-57 (1997).
16 Adam J. Engler *et al.*, Cell 126 (4), 677-689 (2006).
17 H. Hertz, J. reine und angewandte Mathematik 92, 156–171 (1882).
18 F. Rico *et al.*, Phys Rev E. 72 (2), 021914 (2005).
19 Toshihiro Sugitate *et al.*, Curr. Appl. Phys. 9 (4), E291-E293 (2009).
20 AM Gordon, E. Homsher and M. Regnier, Physiol.Rev. 80 (2), 853-924 (2000).
21 K. R. Boheler *et al.*, Circ.Res. 91 (3), 189-201 (2002).
22 P. Young *et al.*, EMBO J. 17 (6), 1614-1624 (1998).
23 ML Costa *et al.*, Brazilian J of Med. and Biol. Res. 37 (12), 1819-1830 (2004).
24 Y. Sugi and J. Lough, Developmental Dynamics 193 (2), 116-124 (1992).
25 MG Frid *et al.*, Dev.Biol. 153 (2), 185-193 (1992).
26 M. Wick, Poult.Sci. 78 (5), 735-742 (1999).
27 AJ Engler *et al.*, J.Cell Biol. 166 (6), 877-887 (2004).
28 RJ Kelm *et al.*, J.Biol.Chem. 269 (48), 30147-30153 (1994).
29 DT Denhardt and M. Noda, J.Cell.Biochem., 92-+ (1998).
30 JB Lian *et al.*, Vitamins and Hormones - Adv. in Res. and App., 55, 443-509 (1999).
31 P. Ducy, V. Geoffroy and G. Karsenty, Connect.Tissue Res. 35 (1-4), 7-14 (1996).
32 L. Hessle *et al.*, Proc.Natl.Acad.Sci.U.S.A. 99 (14), 9445-9449 (2002).
33 M. Sila-Asna *et al.*, Kobe J Med Sci 53 (1-2), 25-35 (2007).

Mater. Res. Soc. Symp. Proc. Vol. 1498 © 2012 Materials Research Society
DOI: 10.1557/opl.2012.1559

Characterizing the effect of substrate stiffness on neural stem cell differentiation

Colleen T. Curley[1], Kristen Fanale[1], and Sabrina S. Jedlicka[1,2,3]

[1]*Bioengineering Program, 111 Research Drive, Bethlehem, PA 18015, USA*
[2]*Materials Science and Engineering Department, 5 E. Packer Avenue, Bethlehem, PA 18015, USA*
[3]*Center for Advanced Materials and Nanotechnology, 5 E. Packer Avenue, Bethlehem, PA 18015, USA*

ABSTRACT

Differentiated neurons (dorsal root ganglia and cortical neurons) have been shown to develop longer neurite extensions on softer materials than stiffer ones, but previous studies do not address the ability of neural stem cells to undergo differentiation as a result of material elasticity. In this study, we investigate neuronal differentiation of C17.2 neural stem cells due to growth on polyacrylamide gels of variable elastic moduli. Neurite growth, synapse formation, and mode of division (asymmetric vs. symmetric) were all assessed to characterize differentiation. C17.2 neural stem cells were seeded onto polyacrylamide gels coated with Type I collagen. The cells were then serum starved over a 14 day period, fixed, and analyzed for biochemical markers of differentiation. For division studies, time-lapse imaging of cells on various substrates was performed during serum withdrawal using the Nikon Biostation. Division events were analyzed using ImageJ to quantify sizes of resulting daughter. Data shows that C17.2 cell differentiation (as dictated by number and type of division events) is dependent upon substrate stiffness, with softer polyacrylamide surfaces (140 Pa) leading to increased populations of neurons and increased neurite length. Our data also indicates that the ability of neural stem cells to express synaptic proteins and develop synapses is dependent upon material elasticity.

INTRODUCTION

Factors contributing to cell fate consist of both chemical and mechanical cues from the microenvironment [1,2]. Recent studies show that the impact of the mechanical properties of the extracellular environment play an important role in regulating many cellular processes, including migration, proliferation, and differentiation [3-8]. Cell response to material elasticity seems to be cell-type specific and correlates to elasticity of a cell's native tissue [8-11]. While previous studies have illustrated the importance of elasticity in regulating stem cell differentiation [7,8], the role of substrate mechanics in directing neuronal differentiation is not clear.

In this study, we explore the effect of substrate stiffness on neurite length, synapse formation, and mode of division during differentiation. Neurite extension and synapse formation are crucial for the development of functional nervous tissue, while mode of division affects the composition of the adult cell population, with symmetric divisions giving rise to two cells with the same fate and asymmetric giving rise to daughters of differing developmental fate[12].

We utilize thin polyacrylamide gels coated with collagen as growth substrates, a widely used method that allows for precise control of material elasticity by varying the ratio of acrylamide to bis-acrylamide in the gel solutions. This yields different degrees of crosslinking in the polymer and a controlled variation of the elastic modulus [11]. We will use C17.2 neural stem cells, gift of Evan Snyder at the Burnham Institute, a multipotent adult neural stem cell line generated via retro-virus-mediated v-myc transfer into murine cerebellar progenitor cells [13]. When transplanted into the adult mouse neocortex, these cells can differentiate into neurons within regions of targeted apoptotic neuronal degeneration [14]. This work explores the role of material properties in guiding differentiation into therapeutically relevant fates for treatment of neurodegenerative diseases.

THEORY

Polyacrylamide Substrate Fabrication

Polyacrylamide gels were fabricated on 22mmx22mm cover glass, as described previously [11,15,16], with slight modification. Briefly, coverslips were flamed, coated with 0.1N-NaOH, and then with 3-aminopropyltrimethoxysilane when dry. Coverslips were washed with ddH$_2$O and then incubated in 70% glutaraldehyde in PBS. The coverslips were washed again with ddH$_2$O and air dried. Polyacrylamide gels were fabricated on the surface of the activated coverslips and particular stiffnesses were achieved by varying the amounts of acrylamide and bis-acrylamide in the gel solution according to the Table III in Johnson et. al 2007. Rain-X coated 18 mm circular cover glass were then placed on top of the gel solution, and the polyacrylamide was allowed to polymerize for 25-60 minutes. The top cover glass was removed, and gels were treated with Sulfo-SANPAH cross-linker solution, rinsed with 50 mM HEPES pH 8, and finally placed in a 0.2 mg/ml Collagen I solution to allow for cell adhesion. Substrates were incubated under UV light overnight to sterilize before cell seeding.

Cell Culture

C17.2 neural stem cells, a gift of Evan Snyder from the Burnham Institute were cultured according to accepted protocol. NSCs were cultured in DMEM high glucose with 10% Fetal Bovine Serum, 5% Horse Serum, and 1% L-Glutamine. Cells were split at less than 1:10 at least once a week. All experiments were performed with cells at passage number 20 or below. Cells were fed 3 times per week by removing half of the old culture media and replacing with an equal amount of fresh media. For the serum withdrawal procedure, cells were fed with serum-free culture media, DMEM high glucose with 1% L-Glutamine. Cells were seeded onto the polyacrylamide gel substrates at a density of 10,000 cells/cm^2 and allowed to grow to about 80% confluency, at which point the serum withdrawal process began. Cells were fixed 14 days after the start of serum withdrawal. A number of synapse samples were cultured for an additional period after this point, during which they were stimulated in high potassium Locke's buffer(95mM NaCl, 50mM KCl, 2.3mM CaCl$_2$, 1mM MgCl$_2$, 3.6mM NaHCO$_3$, 5mM HEPES, 20mM Glucose), either for 15 minutes or 5 minutes, or placed in low potassium Locke's buffer(154mM NaCl, 5.6mM KCl, 2.3mM CaCl$_2$, 1mM MgCl$_2$, 3.6mM NaHCO$_3$, 5mM HEPES, 20mM Glucose) for 15 minutes. Samples were stimulated every 12 hours for a total of 5 days.

Immunocytochemistry

For the neurite growth studies, samples were fixed after 14 days of serum withdrawal and stained for β-tubulin III-AF488 (Covance, A488-435L) at a concentration of 1:100 and rhodamine phalloidin (Cytoskeleton, Inc.) at a concentration of 7 μl per ml. Samples were counterstained with Hoechst dye (Invitrogen) at 0.002 mg/ml in ddH$_2$O for 5 minutes.

For synapse studies, samples were fixed after 14 days of serum withdrawal and stained for synaptophysin (EMD Millipore, MAB368), synaptotagmin (DSHB, mAB 30), homer (Synaptic Systems, 160 103), and PSD-95 (Santa Cruz Biotechnology, sc-71935), all at a concentration of 1:500. The synaptotagmin antibody developed by Louis Reichardt was obtained from the Developmental Studies Hybridoma Bank developed under the auspices of the NICHD and maintained by The University of Iowa, Department of Biology, Iowa City, IA 52242. Secondary antibodies used were Alexa Fluor 546 goat anti-mouse IgG1, Alexa Fluor 488 goat anti-mouse IgG2a, Alexa-Fluor 488 goat anti-rabbit IgG, Alexa Fluor 488 goat anti-mouse IgG2a (Invitrogen), respectively. Images were taken with a Zeiss Observer Z1 inverted fluorescence microscope.

Division Studies

Time-lapse imaging was performed with the Nikon Biostation IM over extended periods of time during the serum withdrawal procedure, with pictures captured every 5 minutes. Mattek dishes(35 mm) were used for cell observation on the biostation, with cells seeded at a density of 2500 cells/cm^2. Dishes were used either as is, coated with a thin 140 Pa gel, or coated with collagen to produce substrates with various stiffnesses and surface properties.

Data Analysis

Neurite growth quantification was performed using the NeuronJ plug-in for ImageJ [18]. Tracings were performed on each visible neurite from the nucleus until the edge of the extension. All images included at least 10 neurite tracings. Daughter cell areas were measured with ImageJ software after each captured division.

DISCUSSION

Cells were fixed and analyzed for protein expression 14 days after the start of serum withdrawal. Neurons were observed on the control substrates (tissue culture treated coverslips), and gels of all tested stiffnesses, as indicated by the expression of β-tubulin III shown in Figure 1. Neurite length was found to decrease with increasing substrate stiffness, with significantly longer extensions observed on the 140 Pa gel. Results are shown in Figure 2. It is also interesting to note that cells seemed to adhere much better to the gel substrates than to tissue culture treated coverslips throughout differentiation and the gels seemed to yield more pure populations of neurons. These results are consistent with previous studies that look at neural cell populations on substrates of various stiffness [10].

Figure 1. Differentiated C17.2 cells stained for β-tubulin III (arrows) and nuclei (asterisks) on: (a.)140 Pa, (b.) 1050 Pa, (c.) 60000 Pa.

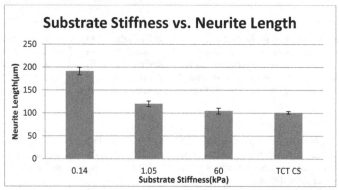

Figure 2. Combination of results from 3 experimental trials. The 1.05 kPa and 60 kPa substrates had neurite length differences to significance level of $p<0.1$, while all other tested substrates had differences to a significance level of $p<0.01$. Calculated using student's t-test.

Since the longest neurite growth occurred on the softest gel tested, we used only 140 Pa gel samples for subsequent analysis of synaptic protein expression after differentiation. We stained for several presynaptic (synaptophysin/synaptotagmin) and postsynaptic (homer/PSD-95) proteins to visualize synapses, which would appear as an area of colocalization between a presynaptic and postsynaptic marker. These samples received additional treatment every 12 hours for five days after differentiation. One set was incubated in low potassium Locke's buffer for 15 minutes to serve as the control, whereas two other sets of samples were incubated in high potassium Locke's buffer for either 5 minutes or 15 minutes. We did not see expression of synaptotagmin or PSD-95 in our C17.2 neurons, but synaptophysin and homer were present and colocalization was observed, as shown in Figure 3. The samples that received the 15 minute treatment in high potassium had more colocalization than the other two groups. This indicates that soft gel substrates are able to support synapse formation of differentiating C17.2 cells and that stimulation in high potassium aides in formation of synapses. High extracellular potassium concentrations depolarize the neurons and induce activity, possibly leading to a more mature neuronal population and increased synapse formation.

Time-lapse images of C17.2 cells on glass, collagen coated glass, and 140 Pa gel substrates were acquired at particular serum concentrations during the serum withdrawal procedure. Daughter cell areas were measured for each observed division and current data can be seen in Figure 4. We are still in the process of compilation and analysis of acquired data.

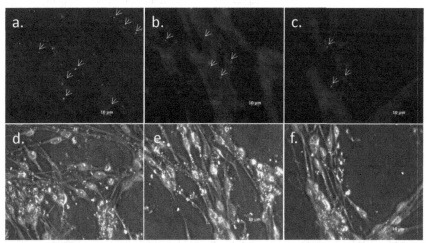

Figure 3. 140 Pa gel samples stained for synaptic proteins, Synaptophysin and Homer. Arrows indicate synapses (colocalization). Stimulation Treatments: (a.)/(d.) 15 min. high K+ Locke's buffer, (b.)/(e.) 5 min. high K+ Locke's buffer, (c.)/(f.) 15 min. low K+ Locke's buffer.

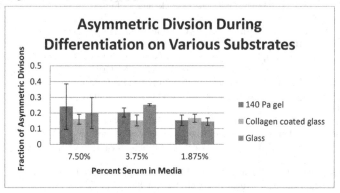

Figure 4. Fraction of asymmetric division events throughout serum withdrawal. Asymmetric events are divisions in which the area ratio of the smaller daughter to the larger daughter is less than 0.7.

CONCLUSIONS

Substrate stiffness affects neuronal differentiation. We have found that our softest substrate, with stiffness comparable to that of brain tissue [10], supports differentiation of C17.2 neural stem cells into neurons with the longest neurite extensions and promotes formation of synapses. To our knowledge, we are the first to characterize synaptic protein expression in this cell type and verify the ability to form synapses *in vitro*. We are also among the first to explore the role of substrate properties on symmetry of division plane during C17.2 differentiation and the effect on developmental fate of the final cell population. This knowledge brings us one step closer to biomaterials scaffolds with tunable mechanical properties for direct control of neuronal fate. Eventually, this technology may lead to materials that can control differentiation of neural stem cells into therapeutically relevant fates for treatment of neurodegenerative diseases.

ACKNOWLEDGMENTS

This research was supported by NSF CBET Grant #1014957 and the Lehigh University 2012 Faculty Innovation grant.

REFERENCES

1. D.E. Discher, D.J. Mooney, P.W. Zandstra., Science **324**, 1673-77 (2009).
2. F. Guilak, D.M. Cohen, B.T. Estes, J.M. Gimble, W. Liedtke, C.S. Chen. Cell Stem Cell **5**, 17-26 (2009).
3. E. Hadjipanayi, V. Mudera, and R.A. Brown., Cell Motil. Cytoskeleton **66**, 121-128 (2009b).
4. C.M. Lo, H.B. Wang, M. Dembo, and Y.L. Wang., J. Biophys. **79**, 144-152 (2000).
5. E. Hadjipanayi, V. Mudera, and R.A. Brown., J. Tissue Eng. Regen. Med. **3**, 77-84 (2009a).
6. J.P. Winer, P.A. Janmey, M.E. McCormick, and M. Funaki., Tiss. Eng. Part A **15**, 147-154 (2009).
7. A.J. Engler, M.A. Griffin, S. Sen, C.G. Bonnetnann, H.L. Sweeney, and D.E. Discher., J. Cell Biol. **166**, 877-887(2004b).
8. A.J. Engler, S. Sen, H.L. Sweeney, D.E. Discher., Cell **126**, 677-689 (2006).
9. T. Yeung, P.C. Georges, L.A. Flanagan, B. Marg, M. Ortiz, M. Funaki, N. Zahir, W. Ming, V. Weaver, P.A. Janmey., Cell motility and the Cytoskeleton **60**, 24-34(2005).
10. L.A. Flanagan, Y.E. Ju, B. Marg, M. Osterfield, P.A. Janmey., Neuroreport **13**, 2411-5 (2002).
11. R.J. Pelham, and Y.L. Wang., PNAS **94**, 13661-5 (1997).
12. B.W. Lu, L. Jan, and Y.N. Jan., Annual Review of Neuroscience **23**, 531-56 (2000).
13. E.Y. Snyder, D.L. Deitcher, C. Walsh, S. Arnold-Aldea, E.A. Hartwieg, C.L. Cepko., Cell **68**, 33-51 (1992).
14. E.Y. Snyder, C.H. Yoon, J.D. Flax, J.D. Macklis., PNAS **94**, 11663-8 (1997).
15. J. D. Aplin and R. C. Hughes., J. Cell. Sci. **50**, 89-103 (1981).
16. K.R. Johnson, J.L. Leight, V.M. Weaver., Methods Cell Biol. **83**, 547-83 (2007).
17. F.X. Jiang, B Yurke, B.L. Firestein, and N.A. Langrana., Annals of Biomedical Engineering **36**, 1565–1579 (2008).
18. E. Meijering, M. Jacob, J.-C. F. Sarria, P. Steiner, H. Hirling, M. Unser., Cytometry **58**, 167-176(2004).

Mater. Res. Soc. Symp. Proc. Vol. 1498 © 2013 Materials Research Society
DOI: 10.1557/opl.2013.11

Hydrogel Composites Containing Carbon Nanobrushes as Tissue Scaffolds

William H. Marks[1], Sze C. Yang[2], George W. Dombi[2] and Sujata K. Bhatia[1]
[1]Harvard University School of Engineering and Applied Sciences,
Cambridge, MA 02138, U.S.A.
[2]University of Rhode Island, Chemistry Department,
Kingston, RI 02881, U.S.A.

ABSTRACT

The objective of this work is to examine the feasibility of electrically conductive hydrogel composites as scaffolds in tissue engineering and tissue regeneration, and to understand the properties of the composites as a growth matrix for clinically relevant cell lines. The composite is comprised of carbon nanobrushes embedded in a biocompatible poloxamer gel. This work assesses the ability of such composite gels to support the growth of fibroblasts and myocytes and eventually serve as a matrix to stimulate wound closure. In such a model, fibroblasts and myocytes are seeded on the hydrogel and bathed in culture medium. The experimental model assesses the ability of fibroblasts and myocytes to grow into and adhere to the gel. The work demonstrates that carbon nanobrushes can be dispersed within poloxamer gels, and that fibroblasts and myocytes can proliferate within homogenously dispersed carbon nanobrush-containing poloxamer gels. This work also examines the effects of carbon nanobrush content on the rheological properties of the poloxamer gel matrix and shows an improvement in several areas in the presence of carbon nanobrushes. Future work will examine the effects of design parameters such as carbon nanobrush content and matrix structure on wound healing, as well as the growth of tendons and other cell lines within the hydrogel composites. This work has relevance for tissue and cellular engineering and tissue regeneration in clinical medicine.

INTRODUCTION

Hydrogels provide three-dimensional encapsulating scaffolds similar to the environment found in vivo in which cells can maintain normal functioning and exhibit tissue growth while being easily examined [1]. Pluronic F-127 poloxamer is a reverse phase change triblock copolymer mixture of polyethylene oxide and polypropylene oxide (PEO_{101}-PPO_{56}-PEO_{101}) that is hydrophilic and non-ionic. The traditional *in vitro* "scar in a jar" model for wound healing, in which two pieces of excised tendon tissue are suspended within a hydrogel and observed for tissue growth between the tendons, utilizes collagen as a hydrogel base; however, poloxamer hydrogel may be a viable replacement for collagen in the "scar in a jar" model, if poloxamer hydrogels can support fibroblast and tendon growth [2].

It has been shown that an alginate gel with embedded carbon nanotubes provided a platform for tissue engineering with a mild inflammatory response and no cytotoxicity [3]; such gels additionally exhibit electrical conductivity [4]. Other studies have also shown that an in vitro collagen matrix provided sufficient support with enough room for fibroblast movement to study adhesion and activity during flexor tendon healing [5]. 3D hydrogels have been produced by electrospinning and have been demonstrated to accurately represent the extracellular matrix [6]. Thus, this study aims to design a suitable hydrogel scaffold for tissue growth using pluronic F-127 poloxamer with electrically conductive carbon nanobrushes (CNB) embedded within the hydrogel. Typical engineered cardiac patches are engineered by seeding cardiac cells within biomaterial scaffolds [7][8], but most of the biomaterials used have poor conductivity and thus limit the patch's ability to contract uniformly [9]. One potential material that allows for improved electrical communication between cells is an alginate scaffold with incorporated gold nanowires bridging the pore walls [10]. F-127 poloxamer with electrically conductive CNB has the potential to overcome those challenges and perform as well or better by improving contractile potential and more accurately replicating tissue. Previous work has shown the ability to integrate and grow cells into the hdyrogel composite [11][12].

EXPERIMENT

Materials and Methods
The CNB were constructed by coating carbon nanotubes with a polyaryl polymer brush. The size of the CNB ranges from 5-20μm in length. The diameter of the CNB is 15-30nm.

The poloxamer solution was prepared as a 30wt% poloxamer solution. First, pluronic F-127 poloxamer (Sigma, USA) was dissolved in 4°C de-ionized water to make a 30wt% solution while continually mixing with a magnetic stir bar and plate. Some manual mixing was necessary if the water warmed up before the poloxamer had dissolved into solution. Then, the mixture was set to rest at 4°C overnight to remove bubbles and fully liquefy the solution before pouring gels.

The poloxamer hydrogels were prepared by pipetting 5mL of solution into each of four wells in a six well plate. The first well was left untouched to prepare a 0vol% CNB gel. In the second, third and fourth wells, a total of 5μL CNB, 25μL CNB and 50μL CNB were added to prepare 0.1vol% CNB, 0.5vol% CNB and 1vol% CNB gels, respectively. Once the hydrogels were prepared, the plate was swirled to uniformly mix the CNB into the hydrogels, and the plate was set to rest in a 4°C fridge for 20 minutes to fully re-liquefy. The plate was then removed and placed in a 37°C incubator until the hydrogels solidified. The solid hydrogel in each well was 5mm in depth. After solidifying, 4mL DMEM (Gibco, USA) and cells were seeded onto the gels in accordance with previous work [11].

Some gels were produced with 0vol% CNB, 0.1vol% CNB, 0.5vol% CNB, 1vol% CNB and 5vol% CNB and not seeded with cells or DMEM. These gels were tested for rheological properties on a TA AR-G2 Rheometer with 20mm steel plate with controlled temperatures. Frequency sweep and time sweep studies were performed. Time sweeps were performed over 10 minute intervals at several points during rheological testing, all at 37°C. Frequency sweeps were performed at an angular frequency of 0.01-5 rad/s at 37°C.

Results
Fibroblast and myocyte density and health were observed by optical microscopy at various layers of the poloxamer hydrogels after 24 and 48 hours. Visualization was performed through a VWR VistaVision Inverted Microscope with TCA 5.0 Color camera attachment.

Primary cardiac fibroblasts were seeded onto 30wt% poloxamer gels containing 0vol%, 0.1vol%, 0.5vol%, and 1vol% CNB. Gels were visualized after 48 hours at 37°C. As the surface of the gel shows (Fig. 1a), the fibroblast cells survived and proliferated in all the CNB gels. The fibroblasts migrated into gels and were also visualized beneath the surface and near the bottom of the gel (Fig. 1b-c).

Primary cardiac myocytes were seeded onto 30wt% poloxamer gels containing 0vol%, 0.1vol%, 0.5vol%, and 1vol% carbon nanobrushes. Following 48 hours incubation at 37°C, the gels were visualized. The surface (Fig. 1d) shows the myocyte cells survived and proliferated in all four gels but did not appear to migrate into the gel nearly as much as the fibroblasts. These results are encouraging in examining potential translations uses for these hydrogel composites as a way to make specific tissue patches or matrices for specialized cells requiring electrical conductivity, including cardiac tissue.

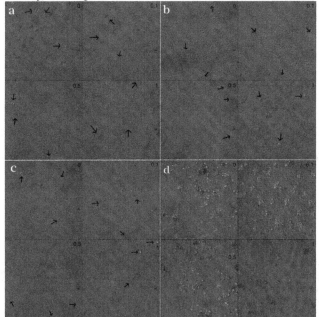

Figure 1. (a) Primary cardiac fibroblasts in top layer of poloxamer gel after 48 hours visualized with light microscopy. (b) Primary cardiac fibroblasts in middle layer of poloxamer gel after 48 hours visualized with light microscopy. (c) Primary cardiac fibroblasts in bottom layer of poloxamer gel after 48 hours visualized with light microscopy. (d) Primary cardiac myocytes in top layer of poloxamer gel after 48 hours visualized by light microscopy. Clusters of myocytes can be seen as brown-beige spots within the gel. Numbers in upper right-hand corner of each image indicate vol% CNB content embedded within poloxamer gel. Arrows point to fibroblasts within that particular layer of gel.

Rheological testing of the hydrogel comoposites containing CNB revealed several interesting properties of the hydrogel and CNB. The storage modulus, G', is a measure of the amount of energy stored by a sample during deformation, and thus available for reformation representing the elastic behavior of the material, while the loss modulus, G", is a measure of the energy used by the sample during deformation and thus lost representing the viscous behavior of the material.

Time sweep studies performed on the hydrogels at multiple times during rheological testing all showed that no polymerization changes were taking place (Fig. 2). This indicates a static state of polymerization at physiological conditions of 37°C for all of the gels from 0vol% CNB to 5vol% CNB, indicating no deterioration of polymeric structure as a result of CNB content.

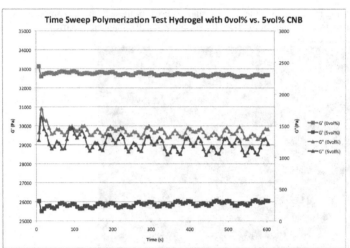

Figure 2. Time sweep polymerization test of poloxamer hydrogel containing 0vol% and 5vol% CNB at physiological conditions (37C).

Frequency sweep studies performed on the hydrogels at 37°C showed a crossover from a regime dominated by elastic effects to one dominated by viscous effects at an angular frequency of approximately 0.03rad/s for the hydrogels, regardless of CNB content (Fig. 3). Increased CNB content led to the regime change at a higher frequency, but the difference was very small. Regardless of CNB content, the hydrogels still had a large G" component at the crossover point, indicating that while viscoelastic effects become dominant, the hydrogels are still quite elastic in nature.

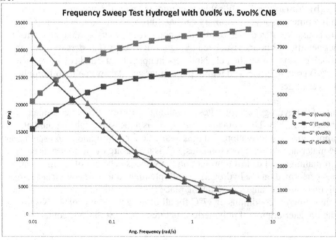

Figure 3. Frequency sweep test of poloxamer hydrogel containing 0vol% and 5vol% CNB at physiological conditions (37C) showing a crossover from predominately elastic to predominately viscous regimes.

DISCUSSION

The presence and proliferation of both primary cardiac fibroblasts and primary cardiac myocytes within the hydrogel composite containing electrically conductive CNB is very promising for applications in short-term and long-term tissue regeneration. The electrically conductive nature of the complex allows for electrically based communication between cells and for clinically relevant connectivity with regard to signaling and synchronized beating. Particularly, the electrically conductive nature of the complex could enhance signal propagation in neuronal tissue and other tissue types, while also enhancing electrical coupling and muscle contractions when used with their respective tissue types. This simulates the electrical connectivity found natively in vivo during tissue development, enhancing the ability to regenerate and repair tissue in situ.

Previous studies on the gelation process of poloxamer 407 at temperatures lower than physiological conditions revealed a mechanism of gelation predicated on micelle packing and entanglement [13]. However, at physiological conditions, the time sweep studies performed indicated that regardless of CNB content, the hydrogels remained completely gelled despite various torques, stresses and strains on the gel throughout testing. The micelle packing and entanglement argument fits well here since the CNB content disrupts that packing and makes the hydrogels with increasing CNB content less elastic than those without CNB. Once gelled, the hydrogels did not depolymerize at all despite CNB content over many months when kept at physiological conditions (37°C).

With respect to G' and G" values, the hydrogels with and without CNB were well within the elastically dominated regime of hydrogels. The frequency sweep showed that there is a crossover from a predominately elastic regime to a predominantly viscous regime at low frequencies. At very low frequencies, near the same time scales as cellular movement, the gel has more viscous properties, likely enabling cells embedded within the gel to remodel the gel as needed for movement and tissue formation. The CNB appear not to noticeably affect this ability. The cells seeded in the hydrogels with higher CNB content showed enhanced movement and proliferation within the hydrogel, compared to those in the 0vol% CNB hydrogel. This suggests that the CNB might in fact affect cellular movement within the gel by functioning in a similar capacity to an extracellular matrix, with structural support and electrical conductivity for enhancing cell signaling.

FUTURE WORK

Future studies will focus on more properties and characteristics of the hydrogel composite as well as looking at the effect of CNB within the composite. CNB made of different types of conducting polymers will be studied, along with other hydrogel composites made from mixtures of polymers. These composites will then be tested for physical properties and use as tissue scaffolds, based on their ability to serve as a matrix for tissue and cell line growth.

REFERENCES
[1] Hunt NC, Grover LM. Cell encapsulation using biopolymer gels for regenerative medicine. Biotechnology Letters 2010; 32(6):733-742.

[2] Chen CZC, Raghunath M. Focus on collagen: in vitro systems to study fibrogenesis and antifibrosis – state of the art. Fibrogenesis Tissue Repair 2009; 2:7.

[3] Kawaguchi M et al. Preparation of carbon nanotube-alginate nanocomposite gel for tissue engineering. Dental Materials Journal 2006; 25(4):719-725.

[4] Basavaraja C, Kim BS, Huh DS. Characterization and AC Electrical Conductivity for the Dispersed Composites Containing Alginate-Multiwalled Carbon Nanotubes. Macromolecular Research 2011; 19(3):233-242.

[5] Packer DL, Dombi GW, Yu PY, Zidel P, Sullivan WG. An in vitro model of fibroblast activity and adhesion formation during flexor tendon healing. The Journal of Hand Surgery 1994; 19(5):769-776.

[6] Coburn J, Gibson M, Bandalini PA, Laird C, Mao HQ, Moroni L, Seliktar D, Elisseeff. Biomimetics of the Extracellular Matrix: An Integrated Three-Dimensional Fiber-Hydrogel Composite for Cartilage Tissue Engineering. Smart Structures and Systems 2011; 7(3):213-222.

[7] Leor J, et al. Bioengineered cardiac grafts: a new approach to repair the infarcted myocardium. Circulation 2000; 102(III):III56-III61.

[8] Dvir T et al. Prevascularization of cardiac patch on the omentum improves its therapeutic outcome. Proceedings of the National Academy of Sciences USA 2009; 106(35):14990-14995.

[9] Bursac N, Loo YH, Leong K, Tung L. Novel anisotropic engineered cardiac tissues: studies of electrical propagation. Biochemical and Biophysical Research Communications 2007; 361(4):847-853.

[10] Dvir T et al. Nanowired three-dimensional cardiac patches. Nature Nanotechnology 2011; 6:720-725.

[11] Marks WH, Yang SC, Dombi GW, Bhatia SK. Translational potential for hydrogel composites containing carbon nanobrushes. 38th Annual Northeast Bioengineering Conference (NEBEC) 2012; 392-393.

[12] Marks WH, Yang SC, Dombi GW, Bhatia SK, "Interactions of Poloxamer Hydrogel Composites Containing Carbon Nanobrushes With Clinically Relevant Cell Lines," *Proceedings of the ASME 2012 Summer Bioengineering Conference*, Fajardo, Puerto Rico, June 2012, SBC2012-80.

[13] Cabana A, Ait-Kadi A, Juhasz J. Study of the Gelation Process of Polyethylene Oxide$_a$-Polypropylene Oxide$_b$-Polyethelene Oxide$_a$ Copolymer (Poloxamer 407) Aqueous Solutions. Journal of Colloid and Interface Science 1997; 190:307-312.

Mater. Res. Soc. Symp. Proc. Vol. 1498 © 2013 Materials Research Society
DOI: 10.1557/opl.2013.181

Novel Biologically Inspired Nanostructured Scaffolds for Directing Chondrogenic Differentiation of Mesenchymal Stem Cells

Benjamin Holmes[1]; Nathan J. Castro[1]; Jian Li[1] and Lijie Grace Zhang[1,2]

[1]Department of Mechanical and Aerospace Engineering and [2]Department of Medicine,

The George Washington University, Washington, DC 20052.

ABSTRACT

Cartilage defects, which are caused by a variety of reasons such as traumatic injuries, osteoarthritis, or osteoporosis, represent common and severe clinical problems. Each year, over 6 million people visit hospitals in the U.S. for various knee, wrist, and ankle problems. As modern medicine advances, new and novel methodologies have been explored and developed in order to solve and improve current medical problems. One of the areas of investigation is tissue engineering [1, 2]. Since cartilage matrix is nanocomposite, the goal of the current work is to use nanomaterials and nanofabrication methods to create novel biologically inspired tissue engineered cartilage scaffolds for facilitating human bone marrow mesenchymal stem cell (MSC) chondrogenesis. For this purpose, through electrospinning techniques, we designed a series of novel 3D biomimetic nanostructured scaffolds based on carbon nanotubes and biocompatible poly(L-lactic acid) (PLLA) polymers. Specifically, a series of electrospun fibrous PLLA scaffolds with controlled fiber dimension and surface nanoporosity were fabricated in this study. In vitro hMSC studies showed that stem cells prefer to attach in the scaffolds with smaller fiber diameter or suitable nanoporous structures. More importantly, our in vitro differentiation results demonstrated that incorporation of the biomimetic carbon nanotubes and poly L-lysine coating can induce GAG and collagen synthesis that is indicative of chondrogenic differentiations of MSCs. Our novel scaffolds also performed better than controls, which make them promising for cartilage tissue engineering applications.

INTRODUCTION

As modern medicine advances, new and novel methodologies have been explored and developed in order to solve and improve current medical problems. One of the areas of investigation that has great promise is tissue engineering [1-3]. For many years, creating polymer scaffolds as a foundation for tissue growth has been widely investigated [4]. A very common and well established method for creating these scaffolds is a process called electrospinning. Electrospinning has been considered favorable because of the ability of researchers to create polymer scaffolds on the micro and nanoscale that mimic the extracellular matrix (ECM) and create an environment which improves cell proliferation and differentiation [5]. And while the system parameters have been thoroughly investigated, research is also being done into how polymer scaffolds can be further modified to increase their physical and compositional complexities, as well as have a greater impact on cell growth. Electrospinning is also favorable because of the ease of scaffold fabrication, the ability to incorporate nano and micro composite material and the ability of scaffolds to influence and promote cell functions.

In a typical electrospinning setup, a high voltage potential is created over a working distance and used to charge polymer chains and draw them to a grounded collector plate [5-7]. A solid polymer is dissolved at a viscous concentration in a volatile solvent. The solvent is then loaded into a syringe with a blunt point needle or capillary and syringe pump. The end of the capillary is then placed at some given distance, usually between 10 and 20 centimeters, from a grounded collector plate and a voltage traditionally between 10 and 20 kilovolts is applied to the capillary. This voltage causes the polymer- solvent solution to be drawn out of the capillary in very fine fibers that accumulate on the grounded collector plate. The solvent then evaporates and a mesh of solid polymer fibers is left over [7-9].

There are two different electrospun scaffold design; 2D and 3D scaffolds. A 2D scaffold is typically only several layers of fibers, which depending on the fibers could be from hundreds of nanometers to under a micron [10]. These scaffolds are also intended to grow only one layer of cells [11]. 3D scaffolds, however, are many layers of fibers thick, and typically are visible to the naked eye. This also allows for cells to be seeded throughout the scaffold, while in a 2D scaffold they can only be seeded on the surface. And while a 2D scaffold has been shown to promote cell growth, 3D scaffolds have been shown to provide much better proliferation and differentiation results [12, 13]. In addition, 3D scaffolds are intended to fully mimic the ECM of natural tissue and have the potential to grow larger amounts of bulk tissue at a time [1, 2, 4].

In native tissue, the ECM forms naturally porous, nanostructured environments that promote cell adhesion, differentiation and proliferation [4, 14, 15]. Therefore, it is important to construct a scaffold that mimics the scale and structure of the ECM [16]. However, it is difficult to get purely electrospun polymers on the truly nano scale, because of the limitations of electrospinning and due to the loss of structural and mechanical integrity of a scaffold with smaller fiber dimensions. Because of these considerations, more and more research has been done to modify the surface characteristics of micro-scale and submicron scaffold fibers. Methods such as co-spinning finer fibers onto thicker support fibers and the incorporation of inherently porous materials have been considered, but an emerging fabrications method that has begun to gain popularity is wet-electrospinning. The process of wet electrospinning entails a standard electrospinning setup augmented with the collector plate submerged in a methanol bath. When the polymer fibers are spun into this bath, they crosslink and form a complex block co-polymeric structure that is highly porous. T. Shin et al wet electrospun a 3D poly(trimethylenecarbonate-co-epsilon-caprolactone)-block-co-poly(p-dioxanone) scaffold for bone regeneration that was 90% porous and exhibited interconnection among pores [17]. This highly porous scaffold showed that cells proliferated 1.5 times faster than the control after seven days [17]. In our study, we will design a novel multi-walled carbon nanotubes (MWCNTs) embedded poly L-lactic acid (PLLA) scaffold with poly L-lysine coating via the wet electrospinning method for directed chondrogenic differentiation of human bone marrow mesenchymal stem cells (MSCs) *in vitro*.

EXPERIMENTS

The purpose of these experiments was to evaluate the effects of nanomaterials on the mechanical properties of electrospun polymer scaffolds and stem cell activities for cartilage repair. It was also a goal to investigate if the nanotubes could be coated with a cell-favorable molecule and effectively delivered to differentiating stem cells, with a positive effect on cell activity. For all experiments, PLLA, purchased from Polysciences Inc. was used as the base polymer to be electrospun. Fibers were fabricated using an in house setup, consisting of a syringe pump, Harvard Apparatus variable voltage supply and an aluminum collector plate.

As-synthesized MWCNTs were obtained from Shanghai Xinxing Chenrong Technology Development CO., LTD. The MWCNT were synthesized by the floating-catalyst technique in chemical vapor deposition process. Dimethylbenzene (C_8H_{10}) and Thiophene (C_4H_4S) served as carbon sources and the Iron atom from Ferrocene ($Fe(C_5H_5)_2$) was used as catalyst for the growth of MWCNT. The synthesis processes were carried out in a cylindrical chamber with temperature of 1100 °C in H_2 environment. The as-synthesized MWCNTs were further treated by the supplier via H_2 heating. H_2 heating treatment involves putting the MWCNTs in a controlled atmospheric environment consisting of a mixture of nitrogen and hydrogen at 800 °C. The MWCNTs are left in this environment for two hours and then the H_2 supply is turned off. The samples are then allowed to cool down at room temperature. The H_2 treatment removes amorphous carbon and nanohorns encapsulating metal catalyst nanoparticles, making tubes more uniform. For the differentiation study, the obtained treated and untreated nanotubes were coated with poly-L-lysine in our lab. Samples were mixed in a solution of deionized water and 0.1% poly-L-lysine and sonicated for 45 minutes, then incubated overnight. The MWCNTs and H_2 treated MWCNTs were imaged using Scanning Electron Microscope (SEM) and Transmission Electron Microscope (TEM).

Primary human bone marrow MSCs were derived from healthy consenting donors from the Texas A&M Health Science Center, Institute for Regenerative Medicine and thoroughly characterized [18]. MSCs (passage #3-6) was cultured in a complete media comprised of Alpha Minimum Essential medium (α-MEM, Gibco, Grand Island, NY) supplemented with 16.5% fetal bovine serum (Atlanta Biologicals, Lawrenceville, GA), 1% (v/v) L-Glutamine (Invitrogen, Carlsbad, CA), and 1% penicillin : streptomycin solution (Invitrogen, Carlsbad, CA) and cultured under standard cell culture conditions (37°C, a humidified, 5% CO_2/95% air environment). For the differentiation study, chondrogenic media was prepared which consisted of the above media recipe, but with the addition of 100nM dexamethasone, 40 µg/ml proline, 100 µg/ml sodium pyruvate, 50 µg/ml L-Ascorbic acid 2-phosphate and ITS+ at a concentration of 1% of the total volume of prepared media.

For MSC adhesion study, pure PLLA was dissolved at 18% weight by volume in a 9 to 1 solution of Dichloromethane (DCM) and Dimethylformaldehyde (DMF). DMF evaporates at a higher temperature than DCM, and was thus postulated to induce surface porosity. A series of microfibrous scaffolds with different diameters were electrospun at 12, 14, 16, 18 and 20 cm working distance from the collector plate, at voltages varying from 14 to 18 kV. Samples of varying working distance were analyzed using SEM. Then MSC adhesion was evaluated on these scaffolds. MSCs were seeded at 10,000 cells per scaffold and cultured in a standard stem cell growth medium for 4 hours. Cells were then counted to evaluate the fiber dimension effects on stem cell adhesion.

For MSC proliferation study, PLLA dissolved in pure DCM at 18% weight by volume, which was then wet-electrospun into a coagulation bath of methanol. Furthermore, PLLA scaffold were electropsun with solutions containing 0.5% w/v untreated MWCNTs, 0.5% w/v H_2 treated MWCNTs and 1% w/v H_2 treated MWCNTs. All samples were homogenously mixed using ultrasonication. The compressive mechanical property of the fabricated scaffolds (8 mm in diameter) was tested using an ATS axial tester. The force-deformation data was then used to calculate and compare the Young's Modulus of each sample. Each sample was also imaged using SEM. In addition, MSCs were seeded at 10,000 cells per scaffold and culture for 1, 3, and 5 days. Cell proliferation was analyzed via a MTS proliferation assay and read via a microplate reader.

Finally, a MSC differentiation study was conducted, using a pure PLLA wet-electrospun scaffold and four experimental groups containing 0.5% w/v MWCNTs, 0.5% w/v H_2 treated MWCNTs, 0.5% w/v MWCNTs coated with poly L-lysine and 0.5% w/v H_2 treated MWCNTs coated in poly L-lysine. The poly L-lysine was used as a means to increase the hydrophilicity of the scaffolds and test the efficacy of MWCNTs dispersed in a polymer scaffold as a chemical delivery device. Samples were cultured in chondrogenic media with MSCs seeded at 250,000 cells per scaffold, and cultured for 1 and 2 weeks. Collected samples were freeze dried in a lyophilizer and treated in a Papain digestion solution, which was then evaluated for glycosaminoglycan content (GAG, DMMB assay) and total collagen content (hydroxyproline assay).

All cellular experiments were run in triplicate and repeated three times for each substrate. Data are presented as the mean value ± standard error of the mean (SEM) and were analyzed with student's t-test for pair-wise comparison. Statistical significance was considered at $p < 0.1$.

RESULTS AND DISCUSSION

MSC Adhesion Study of PLLA Scaffolds with Different Diameters

The MSC adhesion study sought to compare fiber size effects on MSC adhesion, as well as to optimize our setup. As shown in Table I, smaller fiber diameters were yielded by increasing the working distance, with a slight increase at 20 cm. These results can also be observed via SEM imaging (Figure 1). It is also shown that the scaffolds with the

Table I. Electrospun PLLA fibers' diameters, as compared to working distance between the needle and the collector plate.

Working Distance (cm)	12	14	16	18	20
Average Fiber Diameter (µm)	3.390	1.715	1.672	1.332	1.543

smallest fiber diameter promoted the greatest cellular adhesion of MSCs (Figure 2), displaying that the smallest and thus more biomimetic fiber dimensions promote the best adhesion. It should be noted that our larger fibers also have good cell adhesion. This may be due to the fact that the larger fibers had induced nanoscale (60 nm) surface pores, which may have helped to create more surface area for MSC adhesion. This suggested both of fiber dimension and surface topography can contribute to create a biomimetic stem cell-favorable environment. These results are also mirrored in literature. Fang et al fabricated Polycaprolactone (PCL), poly (lactic acid) (PLA) and hydroxyapatite (HA) nanofiber mats via electrospinning for bone tissue regeneration [19]. The average diameters of PCL/PLA nanofiber were in the range of 300-600 nm, while that of PCL/PLA/HA was 154 nm. The effect of nanofiber composition on the osteoblast-like MC3T3-E1 cell adhesion and proliferation were investigated. The MTT assay revealed that PCL/PLA/HA scaffold shows significantly higher cell proliferation than PCL/PLA scaffolds [19].

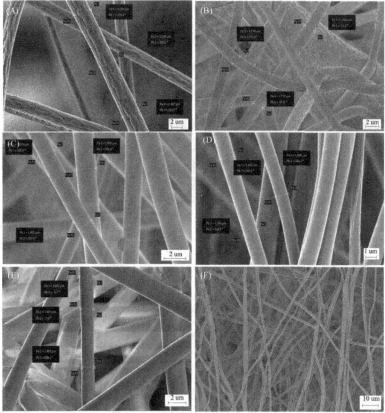

Figure 1. SEM images of electrospun PLLA fibers fabricated at 12, 14, 16, 18 and 20 cm working distance (A-E), and 18 cm at a lower magnification (F).

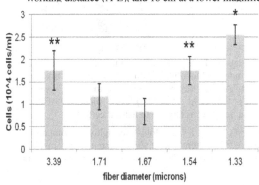

Figure 2. The highest MSC attachment on the electrospun PLLA scaffold with the smallest fiber diameter (1.33 μm, prepared by 18 cm working distance). Data are mean ± SEM, n=9; *p<0.1 when compared to all other samples and **p<0.1 when compared to the scaffold with 1.67 μm fiber diameter.

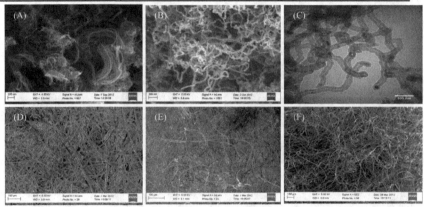

Figure 3. SEM and TEM images of (A) MWCNTs, (B) and (C) H_2 treated MWCNTs. SEM images of (D) pure PLLA scaffold via dry electrospinning, (E) pure PLLA scaffold via wet electrospinning, and (F) 1.0% MWCNTs PLLA scaffold.

Figure 3 shows the SEM and TEM images of the untreated MWCNTs (3A) compared to the H_2 treated MWCNTs (3B and 3C). These images show a change in composition, with the H_2 treated MWCNTs lacking metallic catalyst material. In addition, H_2 heating changed morphology of the nanotube s from bundles of nanotube aggregates into more homogenous distribution, which make them suitable co-electrospin into PLLA scaffold. SEM images were also taken of the electrospun PLLA scaffolds fabricated via dry and wet sp inning and with MWCNTs (Figure 3D-F). The wet electrospun scaffold shows more 3D porous structure than dry electrospun scaffolds. More importantly, after the addition of small concentrations of MWCNTs of both species, the PLLA scaffolds' Young's modulus increased dramatically when compared to a pure PLLA control (Figure 4). And all MWCNT reinforced scaffolds were within the range of native articulate cartilage (~0.75 to 1 MPa) [20-23].

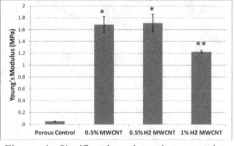

Figure 4. Significantly enhanced compressive Young's Modulus of MWCNT in electrospun PLLA scaffolds when compared to PLLA control (*p<0.05 and **p<0.05). The modulus of the nanocomposites match to natural cartilage.

The proliferation study showed an increase in MSC proliferation on MWCNT scaffolds after three and five days (Figure 5). At five days, all of the nanocomposite scaffolds showed even greater cellular growth, indicating that our MWCNT PLLA scaffolds are highly biocompatible. For the differentiation study, the result showed that there was a significant increase in GAG content (Figure 6) in MWCNT PLLA scaffolds after two weeks. Especially, we observed that the

PLLA scaffolds containing poly L-lysine coated MWCNTs can achieve the highest MSC chondrogenic differentiation. This implies that the surface coating of the nanotubes (to increase hydrophilicity) had the greatest impact in MSC functions. In addition, the positively charged Poly L-lysine can create an electrostatic interaction with negative charged GAG for improved GAG nucleation in the scaffold. Furthermore, the fact that the H_2 treated MWCNTs yielded the better results shows that the purification of nanotubes to remove metal catalyst agents and modify nanotube morphology is advantageous for biological applications.

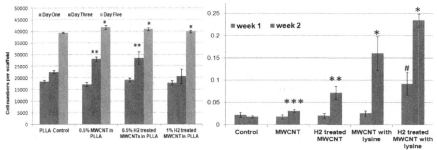

Figure 5. MSC proliferation on various electrospun PLLA / MWCNT scaffolds. Data are mean ±SEM; n = 9. *p<0.05 compared to controls at day 5; **p<0.05 compared to all other scaffolds at day 3.

Figure 6. Significantly improved GAG synthesis of MSCs in all MWCNT (0.5%) embedded in PLLA scaffolds after two weeks. Data are means ± SEM, n=9. #p<0.05 all other samples at week 1; *p<0.05 compared to all other samples at week 2; **p<0.05 compared to control and MWCNTs without H_2 treatment; and ***p<0.05 compared to controls.

CONCLUSIONS

A series of biomimetic electrospun fibrous nano scaffolds with controlled fiber dimension and surface porosity were fabricated in this study. When compared to a pure PLLA control, scaffolds with MWCNTs showed no adverse effect on MSC proliferation, and more importantly displayed enhanced MSC chondrogenic behavior, when modified with a chemical factor on the nanotubes' surface. The purified, H_2 heated nanotubes also showed enhanced GAG synthesis over the untreated tubes, indicating MSCs tending toward chondrocyte lineage on these nanotubes. Scaffolds with incorporated nanotubes also showed significantly improved mechanical properties when compared to a PLLA control. Thus, MWCNTs can be used to greatly improve the strength of electrospun polymer fibers, without having an adverse effect on cellular activity. Overall, the data presented display that an electropsun PLLA scaffold with incorporated, surface modified MWCNTs can be a suitable scaffold for cartilage regeneration.

REFERENCES

1.   R. Langer and J. P. Vacanti, Science **260** (5110), 920-926 (1993).
2.   J. P. Vacanti and R. Langer, Lancet **354 Suppl 1**, SI32-34 (1999).
3.   L. Zhang, J. Hu and K. A. Athanasiou, Crit Rev Biomed Eng **37** (1-2), 1-57 (2009).

4.   D. W. Hutmacher, Biomaterials **21** (24), 2529-2543 (2000).

5.   L. A. Smith and P. X. Ma, Colloids and surfaces. B, Biointerfaces **39** (3), 125-131 (2004).

6.   H. Yoshimoto, Y. M. Shin, H. Terai and J. P. Vacanti, Biomaterials **24** (12), 2077-2082 (2003).

7.   L. S. Nair, S. Bhattacharyya and C. T. Laurencin, Expert opinion on biological therapy **4** (5), 659-668 (2004).

8.   Z. Ma, M. Kotaki, R. Inai and S. Ramakrishna, Tissue engineering **11** (1-2), 101-109 (2005).

9.   A. Thorvaldsson, H. Stenhamre, P. Gatenholm and P. Walkenström, Biomacromolecules **9** (3), 1044-1049 (2008).

10.   X. Meng, W. Li, F. Young, R. Gao, L. Chalmers, M. Zhao and B. Song, Journal of visualized experiments : JoVE (60) (2012).

11.   I. K. Shim, M. R. Jung, K. H. Kim, Y. J. Seol, Y. J. Park, W. H. Park and S. J. Lee, Journal of Biomedical Materials Research Part B: Applied Biomaterials **95B** (1), 150-160 (2010).

12.   L. Chen, C. Zhu, D. Fan, B. Liu, X. Ma, Z. Duan and Y. Zhou, Journal of Biomedical Materials Research Part A **99A** (3), 395-409 (2011).

13.   H. Lee, M. Yeo, S. Ahn, D.-O. Kang, C. H. Jang, H. Lee, G.-M. Park and G. H. Kim, Journal of Biomedical Materials Research Part B: Applied Biomaterials **97B** (2), 263-270 (2011).

14.   M. J. Yaszemski, R. G. Payne, W. C. Hayes, R. S. Langer, T. B. Aufdemorte and A. G. Mikos, Tissue engineering **1** (1), 41-52 (1995).

15.   L. Zhang and T. J. Webster, Nanotoday **4** (1), 66-80 (2009).

16.   M. C. Phipps, W. C. Clem, J. M. Grunda, G. A. Clines and S. L. Bellis, Biomaterials **33** (2), 524-534 (2012).

17.   T. J. Shin, S. Y. Park, H. J. Kim, H. J. Lee and J. H. Youk, Biotechnology letters **32** (6), 877-882 (2010).

18.   D. C. Colter, R. Class, C. M. DiGirolamo and D. J. Prockop, Proc Natl Acad Sci U S A **97** (7), 3213-3218 (2000).

19.   R. Fang, E. Zhang, L. Xu and S. Wei, Journal of Nanoscience and Nanotechnology **10** (11), 7747-7751 (2010).

20.   T. Garg, O. Singh, S. Arora and R. Murthy, Critical reviews in therapeutic drug carrier systems **29** (1), 1-63 (2012).

21.   X. Cui, K. Breitenkamp, M. G. Finn, M. Lotz and D. D. D'Lima, Tissue engineering. Part A **18** (11-12), 1304-1312 (2012).

22.   J. R. Perera, P. D. Gikas and G. Bentley, Annals of the Royal College of Surgeons of England **94** (6), 381-387 (2012).

23.   T. Hogervorst, W. Eilander, J. T. Fikkers and I. Meulenbelt, Clinical orthopaedics and related research (2012).

Mater. Res. Soc. Symp. Proc. Vol. 1498 © 2012 Materials Research Society
DOI: 10.1557/opl.2012.1659

Chemotaxis of Mesenchymal Stem Cells in a Microfluidic Device

Ruth Choa, Manav Mehta, Kangwon Lee, and David Mooney
School of Engineering and Applied Sciences, Harvard University
Cambridge MA, 02138

ABSTRACT

Adult bone marrow derived mesenchymal stem cells (MSCs) represent an important source of cells for tissue regeneration. Control of MSC migration and homing is still unclear. The goal of this study was to identify potent chemoattractants for MSCs and characterize MSC chemotaxis using a microfluidic device as a model system and assay platform. The three chemokines compared in this study were CXCL7, CXCL12, and AMD 3100.

Microfluidic devices made of polydimethysiloxane (PDMS) were fabricated by soft lithography techniques and designed to generate a stable linear chemokine gradient. Cell movements in response to the gradient were captured by timelapse photos and tracked over 24 hours. Chemokine potency was measured via several chemotaxis parameters including: velocity in the direction of interest (V), center of mass (M_{end}), forward migration indice (Y_{FMI}). The migratory paths of the cells were mapped onto a displacement plot and compared.

The following results were measured in the direction of interest (towards higher concentrations of chemokine): For velocity, only cells exposed to CXCL12 had a statistically significant (p=.014) average velocity (V=0.19 ± 0.07 um/min) when compared to the control condition (V=0.06 ± 0.04 um/min). For the center of mass, where the displacement of cells from their starting positions were compared, again only CXCL12 (M_{end}= 53.9 ± 10.8 um) stimulated statistically significant (p = .013) displacement of cells compared to the control condition (M_{end} = 19.3 ± 16.1 um). For the forward migration index, the efficiency of cell movement was measured. Indices in both the CXCL12 (Y_{FMI} = 0.19 ± 0.08) and CXCL7 (Y_{FMI} = 0.09 ± 0.03) conditions were statistically significant (p = .023 for CXCL12 and p = .035 for CXCL7) when compared with the control index (Y_{FIM} = $.04 \pm .02$).

This study demonstrated the use of microfluidic devices as a viable platform for chemotaxis studies. A stable linear chemokine gradient was maintained over a long time scale to obtain cell migration results. CXCL12 was quantitatively determined to be the most potent chemoattractant in this research; these chemoattractive properties promote its use in future developments to control MSC homing.

INTRODUCTION

Bone marrow derived mesenchymal stem cells (MSCs) are multipotent stromal cells that can differentiate into osteoblasts, chondrocytes, and adipocytes.[1] Directly transplanting or intravenously injecting MSCs into wound sites has been shown to be clinically successful in improving wound healing of bony tissues[2]. The migration and homing behavior of these cells can be affected by exposure to various chemokines.

The CXC cytokines are a family of chemokines known to induce chemotaxis in a variety of cell types. There are currently 17 known CXC ligands (CXCL1 through CXCL17) and 7 known CXC chemokine receptors (CXCR1 through CXCR7). These G protein coupled receptors span the membrane and are activated specifically by binding of certain CXC ligands. Two ligands studied in this research are CXCL7, which binds to CXCR 2[3] and CXCL12, which binds to CXCR4.[4] AMD 3100, another ligand in this study, is a small molecule that also binds to CXCR4.[5] While CXCL7 and CXCL12 are known to be chemoattractants for immune cells, recent studies have suggested that they may be chemoattractants for MSCs as well.[6,7]

Gradient generating microfluidic devices have emerged as tools for cellular chemotaxis studies. In these devices, media continuously flows such that the chemokine concentration is kept constant and a steady, well defined gradient can be maintained.[8] These devices are made from Polydimethylsiloxiane (PDMS), which is optically transparent, nontoxic to cells, and can be sealed with glass to form channels. The flow in these channels is laminar, and these devices allow for temporal and spatial control of cellular microenvironments.[9]

EXPERIMENT

MSCs were seeded into the main channel of the microfluidic devices and exposed to chemokine gradients of 0M to 10^{-8}M AMD 3100, 10^{-8}M CXCL7, or 10^{-8}M CXCL12 in 10% FBS DMEM. A control of only 10% FBS DMEM with no gradient was used. Cell movement was tracked over 24 hours with timelapse microscopy and the resulting migratory paths were analyzed with Image J.

Isolation and Culture of Primary Rat MSCs

Lewis rats were sacrificed and bone marrow was collected from the tibia and femur. Cells were centrifuged and resuspended in Dulbecco's modified Eagle's medium (DMEM) with 10% fetal bovine serum (FBS). The cells were seeded in a culture flask at a density of 10,000 cells/cm^2 and cultured under routine conditions of 37 ^{0}C, 10% CO$_2$. The media was changed every 3 days and floating cells were discarded. The adherent cells were cultured to passage 3 and then used in the following experiments.

Microfluidic Device

The microfluidic device (design layout[10]) used in this study is composed of PDMS embedded with a network of microchannels (70 um height) using soft lithography techniques.[11] The PDMS is treated in an oxygen plasma generator (150 mTorr, 100W) for 1 minute and annealed to a glass slide. The micro channel networks generate a gradient of chemokine by splitting, mixing, and combining fluid streams from the two inlets (one with chemokine, the other without). These microchannels are recombined in the main channel (1mm width) such that the concentration gradient is perpendicular to the direction of fluid flow. The gradient (Fig. 1) is maintained throughout the length of the main channel (20 mm) and this main channel acts as an observation chamber where the cells are seeded. The devices are coated with 2% (3-Aminopropyl) triethoxysilane in 200

Figure 1. Gradient Generated in Microfluidic Device
Green dye was injected into the right inlet of the device and water into left. A linear gradient of dye was generated from right to left in the main observation channel.

proof ethanol for optimal cell attachment. The inlet and outlet connections are made with polyethylene tubing and a syringe pump maintains the flow at 1 ul/min.

Cell Seeding, Timelapse Microscopy

Cells were seeded into the microfluidic devices at a density of 10 million cells/ml and allowed to adhere for 4-6 hours. Cells were initially distributed evenly throughout the main observation channel. A chemokine was then added to one of the inlets and cell movement was observed over a period of 24 hours. DIC images were taken every 15 minutes and Image J software was used to track cell movements.

Quantification of chemotactic response of the cells

Due to the variability inherent to cell migration studies, each condition was analyzed at 4-6 different positions, with a more than 100 cells analyzed per condition. Migratory paths of the cells were graphed on displacement maps and chemotactic responses of the cells were quantified with respect to four parameters[12]:

1. **Velocity** V in the direction of increasing chemokine concentration.

$V = \frac{Y}{t}$ *where Y is the displacement of cells in the direction of increasing chemokine concentration and t is the time between photo frames (15 min).*

2. **Center of Mass** M_{end} which represents the averaged displacement of all cells and thus the direction in which cells primarily traveled.

$M_{end} = \frac{1}{n}\sum_{i=1}^{n} y_i$ *where i is the cell index, and y is the displacement in the y direction.n*

3. **Forward Migration Index** Y_{FMI} which represents the efficiency of forward migration in the direction of inter

$Y_{FMI} = \frac{1}{n}\sum_{i=1}^{n} \frac{Y_{i,end}}{d_{i,accum}}$ *where i is the cell index, y_{end} is the end displacement in the y direction, and* d_{accum} *is the total accumulated distance the cell has traveled.*

4. **Directionality** D which is the ratio of the Euclidean to the accumulated distance. It measures how straight the migratory path is.

$D = \frac{1}{n}\sum_{i=1}^{n} D_i = \frac{1}{n}\sum_{i=1}^{n} \frac{d_{i,euclid}}{d_{i,accum}}$ *where i is the cell index, $d_{i,\,euclid}$is the Euclidean distance and* $d_{i,\,accum}$ *is the accumulated distance.*

A Student T test was used to compare the values of these parameters for each chemokine with that of the control condition.

RESULTS

The migratory paths of the cells are shown in Figure 2. The positive Y direction in these plots is the direction of increasing chemokine concentration and thus the direction of interest. The values of the chemotactic parameters for each chemokine are shown in Tables I-IV. Significant p values are bolded.

The chemokine CXCL12 had a significant impact on all four chemotaxis parameters when compared to the control condition. CXCL7 had a significant effect in the forward migration index, and AMD 3100 did not lead to significant effects on any of the parameters. Cells in the control condition (no gradient) did not display preferential migration in any direction

(Fig 2a). The migratory path of cells exposed to AMD 3100 (Fig. 2b) was directed mainly perpendicular to the direction of chemokine gradient rather than parallel to the gradient. Cells exposed to CXCL7 (Fig. 2c) and CXCL12 (Fig. 2d) displayed forward movement in the direction of the chemokine gradient.

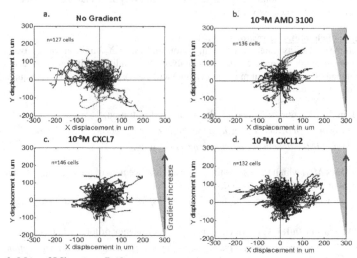

Figure 2. Map of Migratory Paths
(a) the movement of cells with no gradient (10% FBS DMEM in both inlets of the device).
(b-d) cell movement in the presence of a gradient of AMD 3100 (b), CXCL7 (c), and CXCL12 (d). The chemokine gradient runs from bottom to top, with chemokine concentration increasing in the positive Y direction. All plots display X and Y displacement in um from the origin, where the origin is the starting position of cells prior to chemokine exposure.

Chemokine	Average Y direction Velocity in um/min	Standard Error	T-test p value compared to control
CXCL12	0.19	0.02	**.014**
CXCL7	0.07	0.01	.343
AMD 3100	0.06	0.01	.659
Control	0.06	0.02	

Table I. Y direction Velocity

Chemokine	Average Y direction M_{end} in um	Standard Error	T-test p value compared to control
CXCL12	53.9	4.10	**.013**
CXCL7	26.8	4.07	.414
AMD 3100	12.7	2.32	.461
Control	19.3	8.06	

Table II. Center of Mass: $M_{end, y}$

Chemokine	Average	Standard Error	T-test P-value compared to control
CXCL12	0.19	0.03	**.023**
CXCL 7	0.09	0.01	**.035**
AMD 3100	0.13	0.06	.261
Control	0.04	0.01	

Table III. Forward Migration Index Y_{FIM}

Chemokine	Average	Standard Error	T-test P-value compared to control
CXCL12	0.45	0.03	**.001**
CXCL 7	0.25	0.03	.491
AMD 3100	0.30	0.02	.217
Control	0.21	0.04	

Table IV. Directionality D

DISCUSSION AND CONCLUSIONS

A microfluidic device was used to generate a stable, reproducible chemokine gradient.[10] The real time migration of MSCs in response to various chemokines was observed in these devices and quantified statistically. Through an analysis of several chemotaxis parameters, the chemokine CXCL12 was found to be the most potent chemoattractant for MSCs. Although the chemokine AMD 3100 acts on the same receptor (CXCR4) as CXCL12, it was not found to be chemoattractive for MSCs. While CXCL12 was significant in all four chemotactic parameters, CXCL7 also displayed a significant forward migration index. This suggests that CXCL7 may have chemoattractive properties that could become potent at higher chemokine concentrations.

The bone marrow is a rich source of mesenchymal stem cells that participate in the repair and regeneration of bony tissue. Migration of MSCs to a specific wound site occurs early on in the healing process and can critically impact the cascade of healing events that follow.[13] Not only can MSCs eventually differentiate into osteoblasts which lay down new bone, they also secrete other healing factors such as VEGF, bEGF, and IL-6.[14] Control and enhancement of this recruitment stage can lead to further development of methods for bone regeneration.[15]

The CXC family of cytokines has been widely studied for its role in chemotaxis. Its members have been found to regulate the sequential steps of inflammation, angiogeneisis, and tissue repair and regeneration in a variety of tissue types.[16] Recruitment factors for MSCs have only recently begun to be explored though. This study showed that at a concentration of 10^{-8}M, CXCL12 is a MSC chemoattractant. However, CXCL12 is known to be a chemoattractant for numerous other cell types including lymphocytes, endothelial progenitor cells, hematopoietic cells and as well as an inducer of metastasis in some cancer cells.[17] An optimal chemoattractant for MSCs would be both cell specific and noncarcenogenic. For the purposes of bone regeneration, a biomaterial system could be designed to only release CXCL12 locally in the area of a progenitor niche to recruit mainly MSCs.

All chemokines in this study were diluted to 10^{-8}M in full 10% FBS DMEM. The signal to noise ratio may have been improved if a lower percentage of FBS had been used since high serum levels can cause background migration. The 2D environment of the microfluidic assay platform used here is also less physiologically relevant than 3D assay environments. The constitution of a 3D environment may allow for greater cell-cell signaling and could impact homing behavior of MSCs.

REFERENCES

1. Nardi N and da Silva Meirelles L, 2006, Mesenchymal stem cells: isolation, in vitro expansion and characterization, *Stem Cells*, v. 174, 249-82.
2. Wakitani S, Nawata M , et al, 2007, Repair of articular cartilage defects in the patello-femoral joint with autologous bone marrow mesenchymal cell transplantation, *Journal of Tissue Engineering and Regenerative Medicine*, v. 1, 74-9.
3. Kruidenier L, MacDonald TT, et al, 2006, Myofirboblast matrix metalloproteinases activate the neutrophil chemoattractant CXCL7 from intestinal rpithelial cells, *Gastroenterology*, v. 130, 127-36.
4. Balabanian et al, 2005, SDF-1/CXCL12 binds to and signals through the orphan receptor RDC1 in T lymphocytes, *J Biol Chem*, v. 280, 35760-6.
5. Rosenkilde MM, Gerlach LO, Jakobsen JS, et al., 2004, Molecular mechanism of AMD 3100 antagonism in the CXCR4 receptor, *J Biochem*, 279(4):3033-41.
6. Kalwitz G, Endres M, Neumann K, et al., 2009, Gene expression profile of adult human bone marrow derived mesenchymal stem cells stimulated by the chemokine CXCL7, *J Biochem Cell Bio*, 41(3): 649-58
7. Kitaori T, Schwarz EM, Tsutsumi R, et al, 2009, Stromal cell derived factor 1/ CXCR4 signaling is critical for the recruitment of mesenchymal stem cells to the fracture site during skeletal repair in a mouse model., *Arthirits Rheum*, 60(3): 813-23.
8. Sudong K, Hyung JK, Jeon NL, 2010, Biological applications of microfluidic gradient devices, *Integr. Biol*, 2: 584-603.
9. Tang S, and Whitesides G, 2009, Basic microfluidic and soft lithography techniques, *Optofluidics: Fundamentals, Devices, and Applications*.
10. Jeon N, Baskaran H, Dertinger S, et al, 2002, Neutrophil chemotaxis in linear and complex gradients of interleukin-8 formed in a microfabricated device, *Nature Biotechnology*,20(8):836-30.
11. Whitesides G, Ostuni E, Takayama, S, et al, 2001, Soft lithography in biology and biochemistry, *Annu. Rev. Biomed. Eng.* , 3: 335-73.
12. Zengel P, Nyugen-Hoang A, Schildhammer, C, 2011, u-Slide chemotaxis: a new chamber for long term chemotaxis studies, *BMC Cell Biology*, 12(1): 21.
13. McKibbin B,1978, The biology of fracture healing in long bones. *Journal of Bone and Joint Surgery*, 60(2):150-62.
14. Chen L, Tredget EE, Wu PY, Wu Y, 2008, Paracrine factors of mesenchymal stem cells recruit macrophages and endothelial lineage cells and enhance wound healing. *PLoS One*, 3(4):31886.
15. Mehta M et al, 2012, Biomaterial delivery of morphogens to mimic the natural healing cascade in bone. *Adv. Drug Deliv*. Rev.64(12): 1257-76
16. Romagnani P., Lasagni L., Annunziato F., Serio M., Romagnani S., 2004, CXC chemokines: the regulatory link between inflammation and angiogenesis,*Trends Immunol*,25(4):201-9.
17. Kryczek I., Wei S., Keller E., Liu R., Zou W.,2007, Stromal-derived factor (SDF-1/CXCL12) and human tumor pathogenesis., *Am. J. Physiol., Cell Physiol*. 292 (3): C987–95.

Mater. Res. Soc. Symp. Proc. Vol. 1498 © 2013 Materials Research Society
DOI: 10.1557/opl.2013.35

Fructose Enhanced Reduction of Bacterial Growth on Nanorough Surfaces

N. Gozde Durmus[1], Erik N. Taylor[1], Kim M. Kummer[1] and Thomas J. Webster[1,2]

[1] School of Engineering, Brown University, Providence, RI, USA 02912
[2] Department of Chemical Engineering, Northeastern University, Boston, MA, USA 02215

ABSTRACT

Biofilms are a major source of medical device-associated infections, due to their persistent growth and antibiotic resistance. Recent studies have shown that engineering surface nanoroughness has great potential to create antibacterial surfaces. In addition, stimulation of bacterial metabolism increases the efficacy of antibacterial agents to eradicate biofilms. In this study, we combined the antibacterial effects of nanorough topographies with metabolic stimulation (i.e., fructose metabolites) to further decrease bacterial growth on polyvinyl chloride (PVC) surfaces, without using antibiotics. We showed for the first time that the presence of fructose on nanorough PVC surfaces decreased planktonic bacteria growth and biofilm formation after 24 hours. Most importantly, a 60% decrease was observed on nanorough PVC surfaces soaked in a 10 mM fructose solution compared to conventional PVC surfaces. In this manner, this study demonstrated that bacteria growth can be significantly decreased through the combined use of fructose and nanorough surfaces and thus should be further studied for a wide range of antibacterial applications.

INTRODUCTION

Medical devices (such as catheters, renal dialysis shunts, endotracheal tubes) have become an integral part of healthcare. While these devices are used in many areas of medicine for diagnostic and therapeutic purposes, device-associated infections as well as device-associated hospital infections have been a critical concern for many years.[1-4] Contamination of a medical device usually occurs by inoculation with only a few microorganisms from the skin or mucous membranes of the patient during implantation. In addition, the presence of pathogens could be due to in-hospital contamination since they are usually acquired from the hands of the surgical or clinical staff.[5-7] Between 5 and 10% of patients admitted to acute-care hospitals usually acquire one or more infections.[3] Hospital acquired infections (HAIs) are acknowledged world-wide as the most frequent adverse event in health care. HAIs affect approximately 2 million patients each year in the United States, result in up to 100,000 excess deaths, and lead to an estimated cost to the U.S. health care system of more than $35 billion per year.[3, 4] Thus, infection control is critical for patient safety and there exists an urgent need for innovative studies to develop infection prevention strategies in the clinical settings.

Biofilms are a major source of HAIs due to their persistent growth on medical devices and surfaces. For example, patients on mechanical ventilators for extended periods of time often face the risk of developing ventilator associated pneumonia (VAP), an infection in the lung of patients. [8] This is due to the biofilm formation on the endotracheal tubes, thus aspiration of contaminated secretions. [9] To overcome this problem, endotracheal tube surfaces can be coated with antimicrobial agents. However, such coatings may easily delaminate during use. It has been recently shown that creating surface nanotopographies can improve antibacterial properties. For example, changes in the surface nanoroughness and crystallinity can decrease the attachment of different pathogenic strains (*S. aureus*, *S. epidermidis* and *P. aeruginosa*) to a titanium surface.[10] Moreover, it has been shown that the metabolic microenvironment is crucial for antibacterial applications.[11] In this study, our aim was to combine nanorough surfaces with specific sugar metabolites (i.e., fructose) to further decrease bacterial growth and biofilm formation on PVC-based medical devices.

EXPERIMENTAL DETAILS

Commercially available Sheridan® 6.0 mm ID, uncuffed PVC endotracheal tubes (ETT) were cut vertically into 0.6 cm x 0.3 cm segments. A nanorough (NanoR) topography on the ETT surface was created using a lipase from *Rhizopus arrhizus* at a 0.1% concentration in a 1M potassium phosphate buffer (PBS). PVC samples were incubated in the lipase solution at 37°C, 200 rpm for 24 hours. After 24 hours, the lipase solution was replaced with a fresh lipase solution and the samples were incubated for an additional 24 hours. The NanoR and conventional PVC samples were then air dried overnight at room temperature, and they were sterilized using ethylene oxide gas. To combine surfaces with the metabolites, sterilized samples were soaked into 10 mM and 100mM fructose solutions and incubated overnight. The nanotopography of the PVC substrates were evaluated with atomic force microscopy (AFM) under tapping mode using a 9 ± 2 nm AFM tip with scan areas of 5 µm x 5 µm and a scan rate of 0.5 Hz. The root-mean-squared (RMS) roughness was determined for five random fields per sample.[12]

For bacteria studies, *Staphylococcus aureus* (ATCC #25923) was inoculated in 3 mL of tryptic soy broth (TSB) media. The bacteria culture was grown for 18 hours at 37°C, 200 rpm. To mimic an *in vivo* infection, a 10^3 cells/mL solution was seeded into a single well of a 96-well plate containing either a NanoR or conventional PVC sample soaked in 10 mM fructose solution. Unsoaked NanoR and PVC samples were used as controls. After 24 hours, planktonic bacteria growth and biofilm formation on the PVC surfaces were quantified. Planktonic bacteria growth was assessed by the optical density measurements at 562 nm. To assess the biofilm formation, each sample was then placed in 1 ml of PBS and vortexed for 10 minutes. Then, the cells were diluted in phosphate buffer saline (PBS) and plated onto tryptic soy agar (TSA) plates. After overnight incubation, colony forming units were counted to quantify biofilm formation on PVC surfaces.[12]

All experiments were performed in triplicate and were repeated at least three times to validate repeatability and reproducibility. Data were represented as the mean $\pm$ standard error of the mean (SEM). Results were analyzed for statistical significance using a Student's *t*-test (unpaired). Three different significance levels ($p < 0.01$, $p < 0.05$, and $p < 0.1$) were noted.

RESULTS AND DISCUSSION

Material characterization:
SEM analysis indicated that the lipase treatment degraded conventional PVC surfaces and created nanoscale features **(Figure 1)**. In addition, AFM analysis revealed distinct topographies in conventional surfaces compared to NanoR PVC surfaces **(Figure 2)**.

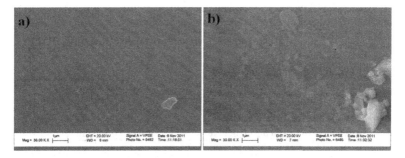

Figure 1. SEM images of untreated PVC and NanoR PVC. **a)** Untreated PVC samples revealed a smooth surface. On the other hand, **b)** NanoR PVC samples revealed a rough surface (65 KX magnification).

The roughness of the conventional PVC surface was 0.704 ± 0.192 nm, while it was 8.440 ± 1.282 nm for the NanoR PVC. In addition, coating with fructose decreased the nanoroughness for both the conventional PVC and NanoR PVC surface. For example, coating the NanoR surface with 10 mM fructose decreased the roughness from 8.440 ± 1.282 nm to 1.214 ± 0.144 nm (Figure 2; Table 1). Surface roughness was significantly higher for NanoR PVC soaked in a 10 mM fructose solution than conventional PVC ($p < 0.01$). When the fructose concentration was increased to 100 mM, RMS values of NanoR decreased from 8.440 ± 1.282 nm to 1.034 ± 0.175 nm. In addition, surface roughness of conventional PVC soaked in 10 mM fructose decreased from 0.704 ± 0.192 nm to 0.592 ± 0.037 nm. In addition, when PVC was soaked in 100 mM fructose, excess sugar deposition increased the roughness to 0.702 ± 0.037 nm.

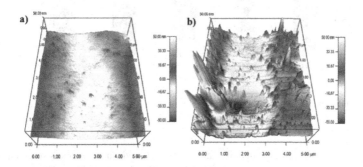

Figure 2. AFM micrographs showing the distinct topography of **a)** control PVC and **b)** NanoR PVC.

Table 1. Summary of RMS values of conventional and NanoR PVC surfaces before and after soaking in different concentrations of a fructose solution.

	Root Mean Square Roughness (RMS)
PVC	0.704 ± 0.192 nm
PVC soaked in 10 mM fructose	0.592 ± 0.037 nm
PVC soaked in 100 mM fructose	0.702 ± 0.029 nm
NanoR	8.440 ± 1.282 nm
NanoR soaked in 10 mM fructose	1.214 ± 0.144 nm
NanoR soaked in 100 mM fructose	1.034 ± 0.175 nm

Assessment of Antibacterial Properties:

Figure 3 shows the planktonic *S. aureus* growth on conventional and NanoR PVC surfaces and the effect of fructose on bacterial growth. The growth of planktonic bacteria decreased on NanoR surfaces soaked in a 100 mM fructose solution compared to conventional PVC surface as well as PVC soaked in respective amounts of fructose. In addition, planktonic bacteria growth decreased on NanoR PVC surfaces soaked in a 100 mM fructose solution compared to NanoR PVC surfaces soaked in a 10 mM fructose solution. Thus, the concentration of fructose metabolites coated on the PVC surface had an effect on planktonic growth.

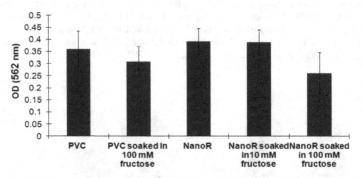

Figure 3. Planktonic *S. aureus* growth on conventional and NanoR PVC soaked in a fructose solution after 24 hours. Data represents mean $\pm$ SEM; N = 3.

Figure 4 shows the biofilm formation of *S. aureus* on PVC and NanoR surfaces soaked in a 10 mM metabolite solution. Metabolic stimulation with fructose metabolites decreased biofilm formation on both PVC and NanoR surfaces after 24 hours. We observed a 38% decrease in the growth of *S. aureus* on PVC surfaces soaked in a 10 mM fructose solution compared to conventional surfaces. In addition, biofilm formation significantly decreased (45%) on NanoR surfaces compared to control PVC surfaces ($p < 0.1$). Moreover, a 60% decrease (~ 0.4 log reduction) in biofilm formation was observed on NanoR surfaces soaked in a 10 mM fructose solution compared to conventional PVC surfaces ($p < 0.05$). More importantly, all these results were accomplished without using any antibiotics.

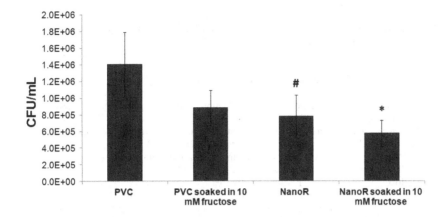

Figure 4. The effect of metabolic stimulation on bacterial growth on nanorough surfaces. Biofilm formation of *S. aureus* biofilm significantly decreased on NanoR surfaces soaked in a 10 mM fructose solution compared to both conventional and NanoR PVC surfaces ($p < 0.05$). Data represents mean ± standard error of the mean (SEM); N = 3.

CONCLUSIONS

There is a clinival need for novel antibacterial surfaces that can resist bacterial growth and biofilm formation. Here, we present for the first time that combining nanorough surfaces with metabolic stimulation (i.e., providing simple fructose metabolites on the surface) can further decrease bacteria functions on PVC-based medical device surfaces. This study also highlights the importance of biofilm metabolic microenvironment for nanomedicine applications. This inexpensive and simple method can be broadly applicable to various medical devices, such as catheters, renal dialysis shunts, and enteral feeding tubes. Development of these novel nanorough surfaces can increase device lifetimes, minimize medical device-associated infections and decrease antibiotic usage in the clinical settings.[12]

ACKNOWLEDGMENTS

The authors would like to thank Hermann Foundation and the Center for Integration of Medicine and Innovation (CIMIT) Prize for Technology in Primary Healthcare for funding.

REFERENCES

[1]   R. M. Donlan, "Biofilms and Device-Associated Infections," *Emerging Infectious Diseases*, vol. 7, 2001.

[2]   Pierce GE, "Pseudomonas aeruginosa, Candida albicans, and device-related nosocomial infections: implications, trends, and potential approaches for control," *J Ind Microbiol Biotechnol.*, vol. 32, pp. 309-318, 2005.

[3]   J. P. Burke, "Infection Control - A Problem for Patient Safety," *N Engl J Med.*, vol. 348, pp. 651-659, 2003.

[4]   Fears R, van der Meer JWM, and Meulen VT, "The Changing Burden of Infectious Disease in Europe," *Science*, vol. 3, pp. 1-4, 2011.

[5]   "The Burden of Health Care-Associated Infection Worldwide: A Summary," *World Healtcare Organization (WHO)*, 2010.

[6]   G. Ducel, J. Fabry, and L. Nicolle, "Prevention of hospital-acquired infections. A practical guide," *World Healtcare Organization (WHO)*, 2002.

[7]   C. Bordi and S. de Bentzmann, "Hacking into bacterial biofilms: a new therapeutic challenge," *Annals of intensive care*, vol. 1, p. 19, 2011 2011.

[8]   Rello J, Torres A, Ricart M, Valles J, Gonzalez J, Artigas A, and R.-R. R., "Ventilator-associated pneumonia by Staphylococcus aureus. Comparison of methicillin-resistant and methicillin-sensitive episodes.," *Am J Respir Crit Care Med.*, vol. 150, pp. 1545-9., 1994.

[9]   Adair CG, Gorman SP, Feron BM, Byers LM, Jones DS, Goldsmith CE, Moore JE, Kerr JR, Curran MD, Hogg G, Webb CH, McCarthy GJ, and M. KR., "Implications of endotracheal tube biofilm for ventilator-associated pneumonia.," *Intensive Care Med.*, vol. 25, pp. 1072-6, 1999.

[10]   Puckett SD, Taylor E, Raimondo T, and Webster TJ., "The relationship between the nanostructure of titanium surfaces and bacterial attachment.," *Biomaterials*, vol. 31, pp. 706-13, 2010.

[11]   Allison KR, Brynildsen MP, and Collins JJ., "Metabolite-enabled eradication of bacterial persisters by aminoglycosides," *Nature*, vol. 73, pp. 216-20, 2011.

[12]   Durmus NG, Taylor EN, Inci F, Kummer KM, Tarquinino KM, and Webster TJ., "Fructose enhanced reduction of bacterial growth on nanorough surfaces.," *Int Journal of Nanomedicine* 2012.

Mater. Res. Soc. Symp. Proc. Vol. 1498 © 2013 Materials Research Society
DOI: 10.1557/opl.2013.241

Bacterial Adhesion to Nanomodified Surfaces:
Dynamic Flow Effects on *S. aureus* and *P. aeruginosa*

Mary C. Machado[1], Keiko M. Tarquinio[2], and Thomas J. Webster[3]
[1]School of Engineering, Brown University, Providence, RI 02912; [2] Pediatric Critical Care
Medicine, Rhode Island Hospital, Providence, RI 02917; [3] Program in Bioengineering and
Department of Chemical Engineering; Boston MA 02115

Abstract

Ventilator associated pneumonia (VAP) is a serious and costly clinical problem.
Specifically, receiving mechanical ventilation over 24 hours increases the risk of VAP and is
associated with high morbidity, mortality and medical costs. Cost effective endotracheal tubes
(ETTs) that are resistant to bacterial infection would help to prevent this problem. The objective
of this study was to determine differences in bacterial growth on nanomodified and unmodified
ETTs under dynamic airway conditions. A bench top model based upon the general design of
Hartmann et al. (1999) was constructed to test of the effectiveness of nanomodified ETTs under
the airflow conditions present in the airway. Twenty-four hour studies performed in a dynamic
flow chamber showed a marked difference in the biofilm formation on different areas of
unmodified tubes. Areas where tubes were curved, such as at the entrance to the mouth and the
connection between the oropharynx and the larynx, seemed to collect the largest amount of
biofilm.

The biofilm formation on ETTs in the airflow system after 24 hours showed a large
difference depending upon where tubes were oriented within the apparatus. This illustrates the
importance of dynamic flow on biofilm formation in pediatric ETTs. It is of particular interest
that increased biofilm density on both unmodified and nanomodified tubes appeared to occur at
curves in the tube where changes in flow pattern occurred. This emphasizes the need for more
accurate models of airflow within pediatric ETTs, suggesting that not only does flow affect
pressure gradients along the tube, but in fact, determines the composition of the film itself. More
testing is needed to determine the effects of biofilm formation on the efficiency of ETT under
airflow, however this study provides significant evidence for nanomodification alone (without
the use of antibiotics) to decrease bacteria function.

Introduction

Ventilator associated pneumonia (VAP) is one of the most common causes of hospital
associated infection in children and adults. *Pseudomonas aeruginosa* (*P. aeruginosa*) and
Staphylococcus aureus (*S. aureus*) are two common strains associated with VAP. However,
multi-drug resistant strains of bacteria such as Methicillin-resistant *Staphylococcus aureus*
(MRSA), are of particular concern to clinicians because of their increasing prevalence within
hospitals. Eight to 28% of all patients receiving mechanical ventilation will develop VAP.
Depending on the pathogen involved, the patient's underlying condition, and the length of
intubation, VAP can have a high mortality rate, ranging between 38% and 76%. VAP can also
be very costly for both patients and hospitals. This condition adds an estimated $40,000 dollars
to each hospital admission due to the extra intensive care required for treatment [1].

Pediatrics creates a challenging environment for the diagnosis of VAP. Radiographic
and clinical criteria for such diagnoses are often unspecific. This causes delays in targeted
treatment and overuse of a broad-spectrum of antibiotics. Specifically, tracheal colonization must

be differentiated from lower respiratory infection. Cultures taken from patients with suspected pneumonia often yield false positives detecting benign bacteria colonization or false negatives missing the active area of bacteria infection. More than half of patients diagnosed with clinical VAP have negative cultures [2]. The diagnosis of VAP in pediatric patients is hindered further by the lack of VAP studies in children and infants.

Endotracheal tubes, one of the main sources of bacterial colonization in the airway, are essential to the process of mechanical ventilation. Like any other device implanted within the body, ETTs interact with an environment that contains not only the airway epithelium, but also potentially harmful microorganisms. Endotracheal tubes present a special concern to clinicians because they are often colonized by oropharyngeal bacteria during long-term mechanical ventilation. These tubes provide a conduit from the outside environment to the more sterile area of the lungs by impairing the body's natural defenses. Moreover, injury to the epithelial cells of the trachea can result from the movement of the tube in the airway or even from the suction of secretions during nursing care. This causes "opportunistic adherence" of bacteria to the airway.

One of the key challenges towards the inhibition of bacterial growth on ETTs occurs when certain bacteria exude an exopolysaccharide, which adheres bacteria together. Bacteria in this type of extracellular matrix (known as a biofilm) are especially resistant to both antibiotics and the immune system of the patient. Ventilation through these infected tubes, or even condensation of humidified air can break off pieces of this biofilm bringing bacteria deeper into the lungs and spurring growth on other areas of the tube [3].

In the past, efforts to reduce VAP have concentrated on decreasing bacterial contamination during intubation by modifying medical procedures. It has become increasingly apparent that the elimination of bacteria also depends on the reduction of bacterial growth on the ETTs within the body.

Cost effective ETTs that are resistant to bacterial infection provide an essential tool to prevent VAP. Nanomodified devices used within the body mimic their natural surroundings and subsequently often lack detrimental immune responses to the foreign object. Research shows a connection between nanophase surfaces (that is, surfaces with features less than 100 nm in at least one direction) and altered bacterial metabolism suggesting the topography of nanostructures influences the biological processes of bacteria [4]. The addition of nanomaterials to these bacteria modifying surfaces could enhance their anti-microbial properties. Preliminary research has suggested that nanoroughened PVC could have antimicrobial properties, and could possibly interfere with bacterial adhesion.

However, such preliminary studies were performed under static conditions and do not take into account the dynamic forces that occur in the human airway. Our model seeks to characterize the antimicrobial properties of nanomodified ETTs within a continuously contaminated airway model. This study hypothesizes that airflow and the impact of continuous bacterial contamination on ETTs are essential to the determination of the bacterial resistance of these tubes in the pediatric airway.

Materials and Methods

Polyvinyl chloride (PVC) ETTs with nano-roughened surfaces were created using lipases from the fungi *Candida cilindracea* and *Rhizopus arrhisus* (Sigma Aldrich) which enzymatically degraded the PVC material. PVC ETT (Sheridan®) were exposed to a 0.1% mass solution of either *C. cilindracea* or *R. arrhisus* lipase dissolved in a potassium phosphate buffer at 37° C. These samples were then gently agitated for 24 hours at which point they were washed with distilled water and exposed to a fresh enzyme solution for another 24 hours. The tubes remained in contact with the lipase solution for a total of 48 hours. The activity of the *R. arrhisus* used in the experiment was 10.5 U/g where one unit was defined as the amount of enzyme that catalyzed the release of 1 μmol of oleic acid per minute at pH 7.4 and a temperature of 40°C.

To better characterize the surface of the nanomodified samples, these samples were analyzed using scanning electron microscopy (SEM). Each of the samples was coated with a gold palladium mixture to increase conductivity and analyzed with a LEO 1530VP FE-4800 Field-Emission SEM, Carl Zeiss SMT, Peabody, MA.

Static studies were performed to analyze bacteria commonly found in VAP. *S. aureus* (ATCC #25923) and *P. aeruginosa* (ATCC #25668). were inoculated into trypticase soy broth (TSB) media. Polyvinyl chloride (PVC) pieces were then immersed into the inoculated media and into a control containing media without bacteria. Bacterial growth on the surface of the PVC was assessed at 4, 12, 24, and 48-hour time points. The bacteria found on these samples were stained with crystal violet and quantified using optical density.

The ultimate test of the effectiveness of nanomodified ETTs under the conditions present within the airway, resulted in the construction of a bench top model. Hartmann et al. (1999) created a continuously contaminated airway system to test the effectiveness of silver coated endotracheal tubes (SCET). The airway model for this study was based upon this general design and is shown in Figure 1 [5].

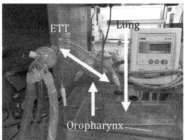

Figure 1. Endotracheal Tube System

Nanomodified ETTs were tested in a custom-made bench top model airway. ETTs with a 3.5 mm internal diameter (ID) were placed in the system. The boxes were then filled with a total of 500 mL of sterile trypticase soy broth media (TSB) and placed in a water bath that was maintained at 37°C. An ETT was connected to an Infant Star 950 ventilator (Puritan Bennett, Covidien, Mansfield, MA) with positive end-expiratory pressure (PEEP) of 1 cmH₂O and a fraction of inspired oxygen (FiO₂) of 0.5. Four hundred and eighty mL of TSB was inoculated with 10^3 colony forming units/milliliter (CFU/ mL) of *S. aureus* (ATCC #25923) or *P. arrugenosa* (ATCC #25668). This bacterial media was then circulated into the oropharynx box over the duration of a 24-hour test, using a peristaltic pump. Sterilization tests were performed routinely to detect any cross contamination. For each run, an ETT was added to the sterilized oropharnynx and lung boxes.

Additionally, at the end of each trial, ETTs were cut into ten 1.5 cm pieces. The inside of each tube was scraped with a sterile polyester applicator (Puritan Bennett, Covidien, Mansfield,

MA). These applicators were then processed using a vortex protocol to determine the number of colony forming units.

All experiments in the system were performed three times ($N = 3$) and the results reported as mean ± standard error of the mean. ANOVA statistical analysis was performed on all results.

Results

The treatment of PVC tubes in either lipase solution (*C. cilindracea* or *R. arrhisus*) showed visible nanofeatures on the ETT tube surface. These nanofeatures can be seen in Figure 2. Untreated PVC did not show any nanofeatures.

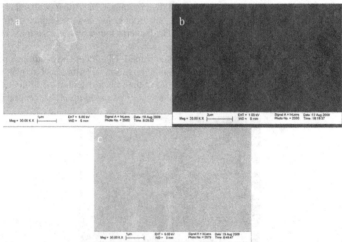

Figure 2. . SEM images of nanomodified PVC ETTs: (a) Nano-C: Mag. x 30K, (b) Nano-R: Mag. x 20K, and (c) Control: Mag. x 30K. Nano-R was modified with *R. arrhisus* while Nano-C was modified with *C. cilindracea.*

Results showed that nanomodified PVC ETTs were effective at reducing bacterial colonization in static studies (Figure 3). *S. aureus* was reduced at the 24 and 48 hr time point within these static studies.

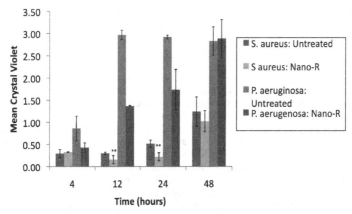

Figure 3. Static *S. aureus* and *P. aeruginosa* mean crystal violet staining on nano-structured PVC ETT, N=3; Data=Mean +/- 1 SE, **$p < 0.01$ compared to controls at same time points.

Dynamic lung system tests on both untreated and nanomodified tubes can be seen for *S. aureus* in Figure 4. Colony forming units are plotted over the length of the tube. Results from this analysis of only the inner surface of the tube still demonstrated uneven bacterial growth over the tube. This was correlated to oscillatory shear stress and particle residence time using an associated finite element model.

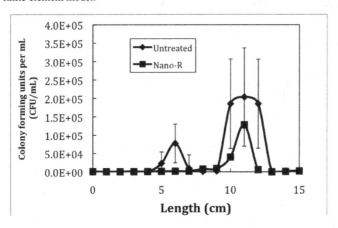

Figure 4. Dynamic *S. aureus* lung system vortexing results for nano-structured PVC ETT, N=3; Data=Mean +/- 1 SE, **$p < 0.01$ compared to controls at same time points.

Discussion

Previous studies of pediatric ETTs have used static bacterial studies to test ETT effectiveness in the airway, without considering the ventilation or dynamic contamination that occurs *in vivo*. This study aimed to simulate these important attributes of the human airway and to use this knowledge to better quantify the antimicrobial properties of ETTs. The data obtained from the dynamic lung system, suggests that airflow and continuous contamination are important in biofilm formation and ETT resistance. Data isolated from the inner lumen of the ETT also suggests that early bacterial colonization and biofilm formation along the tube is greatly influenced by airflow. The 3.0 and 3.5 millimeter tubes used within the dynamic airway system have been shown to exhibit a laminar flow regime [6]. Differences in the air flow around the curves in these tubes either caused by physiological or medical treatment, could play a key role in determining the sights of early bacterial colonization.

Moreover, the static studies on the anti-microbial nature of the nanomodified ETT have shown significant reductions in growth at the 12 and 24-hour time points. Preliminary studies suggest that bacterial number on these tubes is lower than that of untreated tubes. This may suggest that nanoroughness has an impact on not only bacterial number but biofilm composition. By reducing early colonization or changing the distribution of colonization along the ETT, nanomodified ETT could have a great impact on many of the problems caused by bacterial colonization and biofilm formation such as increased resistance to flow and further contamination of the patient. Further investigation into the wall shear stress of nanomodified tubes and the influence of fluid effects on these surfaces is warranted.

Conclusions

These *in vitro* studies have shown that chemical etching with bacterial lipases can create nano-rough surface features on PVC in an inexpensive but effective manner. Static studies have also demonstrated that these nano-roughed surfaces can inhibit *S. aureus* growth.

The exposure of ETT to dynamic lung conditions has effected both the concentration and location of bacterial growth on the ETT. These results suggest that these nanomodified tubes could provide clinicians with an effective tool to combat hospital acquired infections like VAP and should be studied in greater depth.

Acknowledgements

The authors of this paper would like to thank the Rhode Island Science and Technology Advisory Committee for funding this project.

References

1. Office of Quality and Performance, *FY 2008, Q1 technical manual for the VHA performance measurement system*, 315 (2007).
2. F. K. Bahrani-Mougeot, B. J. Paster and S. Coleman, "Molecular analysis of oral and respiratory bacterial species associated with ventilator-associated pneumonia." *J Clin Microbiology* 45, 1588-1593 (2007).
3. J. J. Marini and A. S. Slutsky, *Physiological basis of ventilatory support*. New York: Marcel Dekker (1998).
4. P. Cardinal, P. Jessamine, C. Carter-Snell, S. Morrison, and G. Jones, "Contribution of water condensation in endotracheal tubes to contamination of the lungs." *Chest*, 127-129 (1993).
5. M. Hartmann, J. Guttmann, B. Muller, T. Hallmann, and K. Geiger, "Reduction of the bacterial load by the silver-coated endotracheal tube (SCET) a laboratory investigation." *Technology and Health Care*, 359-70 (1999).
6. Guttmann J, Eberhard L, Fabry B, Bertschmann W, Wolff G. "Continuous calculation of intratracheal pressure in tracheally intubated patients." *Anesthesiology* 79, 503–13 (1993)

Mater. Res. Soc. Symp. Proc. Vol. 1498 © 2013 Materials Research Society
DOI: 10.1557/opl.2013.242

The Effect of Cellulose Nanofibres on Mechanical Properties and Bioactivity of Natural Polymers

Ali Negahi Shirazi, Ali Fathi and Fariba Dehghani
School of Chemical & Biomolecular Engineering, The University of Sydney, NSW 2006, Australia.

ABSTRACT

Natural polymers, used for hydrogel fabrication, are generally bioactive and provide good environment for cell growth and proliferation. However, these polymers have low mechanical strength. Several approaches have been attempted to improve their mechanical properties such as fabrication of interpenetrating polymer network (IPN) and semi-IPN hydrogels, and also addition of a nano sized fibers or nano-particles. The aim of this study was to investigate the feasibility of using naturally derived nano-fillers such as cellulose nanocrystallines to enhance the mechanical properties of hydrogels. Gelatin methacrylate (GelMA) was used as a protein model for preparation of photo-crosslinked hydrogel. The effects of concentrations of photo initiator and cellulose nanocrystallines (CNC) on the characteristics of hydrogels were examined. *In vitro* studies showed negligible cytotoxic effect of CNC on human osteosarcoma cell growth when using less than 20 mg/ml CNC. Therefore, it is viable to use this nano-filler for biomedical applications. It was found that the compression modulus of gelatin hydrogel was increased 1.5 fold by addition of 10 mg/ml of CNC. These results demonstrate the high potential of using CNC for tissue engineering applications to enhance the mechanical strength of hydrogels.

INTRODUCTION

Gelatin is an irreversibly hydrated form of collagen broadly used in tissue engineering [1]. This biocompatible hydrogel possesses poor mechanical properties in physically crosslinked form. Chemically crosslinking of gelatin enhanced its mechanical performances. The cytotoxic nature of these crosslinking agent, however, restricts their biomedical application [2] Functionalization of amine-containing side groups of gelatin upon methacrylate groups resulted to a light crosslinkable hydrogel [3]. This stable hydrogel showed an improved mechanical performance in comparison with physically crosslinked gelatin [3]. Even though photocrosslinking is a rapid technique to fabricate gelatin methacrylate (GelMA) hydrogels, the mechanical properties of these hydrogels are relatively poor. Therefore, incorporation of reinforcing materials to GelMA hydrogels is inevitable. For instance, incorporation of 5 mg/ml silk to GelMA hydrogel significantly improved the compression modulus of hydrogel fivefold [4]. However, the inflammatory responses of silk might limit its applications as a reinforcing material. To overwhelm this drawback, the low immunoresponsive materials such as cellulose can be used as a reinforcement agent [5, 6]. Cellulose nanocrystallines (CNC) synthesized upon acid hydrolysis of cellulose had a modulus around 100 GPa [7]. The purpose of this study is to investigate the feasibility of using CNC as a cost effective reinforcing agent for improving the mechanical performance of hydrogels.

EXPERIMENT

Materials

Gelatin and methacrylate anhydride were supplied by Sigma-Aldrich (St. Louis, USA) and used as received. Eucalyptus derived cellulose sheets were kindly provided by the Federal University of Viçosa, Brazil and used without further purification. Sulphuric acid and all other solvents and reagents were purchased from Sigma-Aldrich (St. Louis, USA) and used as received.

Preparation of CNC

CNC was prepared upon hydrolysis of eucalyptus pulps. The ground cellulose was dissolved into 64 wt% sulphuric acid at a ratio of 1:17.5 (g:ml). The hydrolysis was performed at 45°C for one hour, followed by 10 times dilution in deionized water to cease the reaction. The residue of acid was then removed by several centrifugation cycles. The suspension was further purified by dialysis against deionised water until approaching pH 6. Afterwards, the suspension was ultra-sonicated and vacuum filtered using 0.2 μm filter paper.

Cytotoxicity Assay

Human osteosarcoma cell lines (MG36) were cultured in Dulbecco's modified Eagle's medium (DMEM), supplemented with 10% fetal bovine serum (FBS) and 1% Anti-Anti reagent (Invitrogen) at 37°C in humidified atmosphere with 5% CO_2. Different concentrations of pre-neutralized CNC suspension were added in each well with density of 8×10^4 cell/cm^3. Cytotoxicity of CNC was determined using 3-(4,5-dimethylthiazol-2-yl)-5-(3-carboxymethoxyphenyl)-2-(4-sulfophenyl)-2H-tetrazolium (MTS) assay 7 days post-seeding. 50 μL of MTS reagent (Promega CellTiter 96® AQueous Non-Radioactive, Madison, USA) were added to different concentration of CNC in combination with 250 μL fresh media. Following 1 h incubation at 37°C, the numbers of viable cells were determined upon absorbance measurement of soluble formazan at 490 nm. Values obtained without any cells were considered as background and tissue culture polystyrene (TCPS) was chosen as control. Cell viability was calculated using Equation 1.

$$Cell \ \ Viability = \frac{Absorbance \ \ of \ \ sample \ \ at \ \ 490 \ \ nm}{Absorbance \ \ of \ \ TCPS \ \ at \ \ 490 \ \ nm} \times 100 \qquad \text{Equation 1}$$

Synthesis of GelMA

Gelatin was dissolved in phosphate buffered saline (PBS) at 50°C with concentration of 100 mg/ml. Methacrylic anhydride was added to the gelatin solution at 0.5 ml/min. It was allowed to react for one hour under constant stirring. The solution was 5 times diluted with preheated PBS at 50°C to cease the reaction. The solution was dialyzed against distilled water using 12-14 kDa dialysis tubes for 1 week at 37°C to remove salts and residues of methacrylic acid. Subsequently, it was lyophilized for 3 days, forming white porous foam and stored at -80°C.

Fabrication of Gelatin-based Composite hydrogels

GelMA hydrogel was fabricated by mixing GelMA (100 mg/ml), CNC suspension (0.5-2 % (wt/v)) and 2- hydroxy-1-(4-(hydroxyethoxy)phenyl)-2-methyl-1-propanone (Irgacure 2959®) as a photo-initiator. Three-dimensional hydrogels were formed after crosslinking using UV light at 365 nm and 6.9 mW/cm^{-2} for a period of 2 minutes. The effects of CNC and Irgacure 2959® concentrations on physicochemical properties of hydrogels were investigated.

Mechanical Properties

Uniaxial compression tests were performed in an unconfined state using Instron mechanical testing and 100 N load cell. Prior to mechanical testing, disks of hydrogels (Φ= 6 mm, h=6 mm) were swelled for 2 h in PBS. The compressive properties of the samples were measured triple in the hydrated state, at 37°C. The tangent slope of the stress–strain curve in the linear strain range (10-20%) was used to measure the compressive property.

Swelling Properties

The swelling behavior of porous hydrogel was measured at 37°C in PBS (pH 7.2–7.4). After immersion in excessive PBS at 37°C overnight (12 h), the swollen hydrogels were weighed (Ws). The hydrogels were subsequently lyophilized overnight, and the dry weights were recorded (Wd). The equilibrium swelling ratio (ESR) was then calculated based on following Equation 2.

$$ESR(^{wt}\!/_{wt}) = \frac{Ws - Wd}{Wd}$$

Equation 2

The data presented is based on at least 3 samples.

Statistical Analysis

Data is reported as mean ± STD. One-way analysis of variance (ANOVA) for single comparisons and Bonferroni post hoc tests for multiple comparisons were performed, using IBM SPSS software for Windows, version 19.0.1. Statistical significance was accepted at $p<0.05$ and indicated in the Figures as * ($p<0.05$), ** ($p<0.01$) and *** ($p<0.001$) considering that no star represents statistically insignificant variation.

RESULTS AND DISCUSSION

Cytotoxity of CNC

The cytotoxicity of CNC at different concentrations was determined using MTS assay as shown in Figure 1. CNC with concentration of 5 mg/ml significantly promoted cell proliferation and viability ($p<0.001$). The cell numbers was decreased to 62% of pre-cultured cells when 20 mg/ml suspension of CNC was added to culturing media. This was still higher than half maximal inhibitory concentration (IC50) value. Thus CNC had negligible cytotoxic effect within the range examined. The effect of this nano-filler on physical properties of photocrosslinked GelMA was examined by using 10 mg/ml CNC.

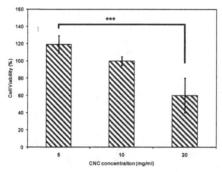

Figure 1. Cell viability on CNC after 7 days. Student's t-tests were performed to demonstrate difference between two groups (*** $p < 0.001$).

Mechanical Properties of GelMA hydrogel

Addition of 10 mg/ml of CNC significantly ($p<0.05$) enhanced the mechanical strength of GelMA hydrogel, shown in Figure 2-a.

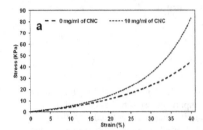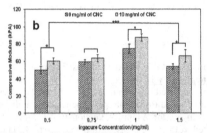

Figure 2. Stress/strain curve of neat GelMA and GelMA hydrogel with 10 mg/ml CNC (a) and compression modulus of GelMA hydrogels, fabricated with different Irgacure and CNC concentrations (b), data reported as *<0.05 and *** p < 0.001.

Addition of hydrophobic CNC into GelMA hydrogel had no significant impact on the swelling behavior of hydrogels (p>0.05) [5]. Results in Figure 3 show that the swelling properties of hydrogels, however, significantly ($p<0.05$) decreased by elevating the concentration of photo-initiator (Irgacure), due to higher level of crosslinking.

Swelling Properties of GelMA hydrogel

Addition of CNC into GelMA hydrogel had no significant effects on the swelling behavior of hydrogels (p>0.05) besides the hydrophobic nature of cellulose [5]. Results in Figure 3 show that the swelling properties of hydrogels however significantly ($p<0.05$) decreased by elevating the concentration of photo-initiator (Irgacure). This effect was due to higher level of crosslinking in hydrogels with elevated concentration of photo-initiator.

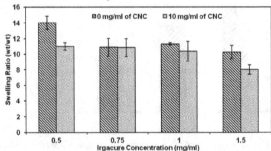

Figure 3. Swelling ratio of hydrogels fabricated upon various concentration of Irgacure 2959® and incorporation of CNC.

CONCLUSIONS

The aim of this study was to investigate the feasibility of using naturally derived reinforcing particles to enhance the mechanical properties of natural-based hydrogels. Gelatin was used as a model protein and functionalized with methacrylate groups to prepare photo-crosslinkable hydrogels. Cellulose nanocrystallines were synthesized with acid hydrolysis of cellulose pulps. *In vitro* studies showed negligible cytotoxic effect of CNC on human osteosarcoma cell growth. Incorporation of CNC into hydrogels enhanced mechanical performance of gelatin-based composites. The addition of CNC had also no significant impact on swelling properties of hydrogels. These results demonstrated the high potential of using CNC to enhance the mechanical strength of hydrogels for different tissue engineering applications.

REFERENCES

1. Benton, J.A., C.A. Deforest, V. Vivekanandan, and K.S. Anseth, *Photocrosslinking of gelatin macromers to synthesize porous hydrogels that promote valvular interstitial cell function.* Tissue Engineering - Part A, 2009. **15**(11): p. 3221-3230.
2. Mathew, A.P., K. Oksman, D. Pierron, and M.-F. Harmand, *Fibrous cellulose nanocomposite scaffolds prepared by partial dissolution for potential use as ligament or tendon substitutes.* Carbohydrate Polymers, 2012. **87**(3): p. 2291-2298.
3. Nichol, J.W., S.T. Koshy, H. Bae, C.M. Hwang, S. Yamanlar, and A. Khademhosseini, *Cell-laden microengineered gelatin methacrylate hydrogels.* Biomaterials, 2010. **31**(21): p. 5536-5544.
4. Xiao, W., J. He, J.W. Nichol, L. Wang, C.B. Hutson, B. Wang, Y. Du, H. Fan, and A. Khademhosseini, *Synthesis and characterization of photocrosslinkable gelatin and silk fibroin interpenetrating polymer network hydrogels.* Acta Biomaterialia, 2011. **7**(6): p. 2384-2393.
5. Eichhorn, S.J., *Cellulose nanowhiskers: promising materials for advanced applications.* Soft Matter, 2011. **7**(2): p. 303-315.
6. Dong, H., K.E. Strawhecker, J.F. Snyder, J.A. Orlicki, R.S. Reiner, and A.W. Rudie, *Cellulose nanocrystals as a reinforcing material for electrospun poly(methyl methacrylate) fibers: Formation, properties and nanomechanical characterization.* Carbohydrate Polymers, 2012. **87**(4): p. 2488-2495.
7. Rusli, R. and S.J. Eichhorn, *Determination of the stiffness of cellulose nanowhiskers and the fiber-matrix interface in a nanocomposite using Raman spectroscopy.* Applied Physics Letters, 2008. **93**(3): p. 033111-3.

Mater. Res. Soc. Symp. Proc. Vol. 1498 © 2013 Materials Research Society
DOI: 10.1557/opl.2013.331

Three Waves of Disinfectants to Inactivate Bacteria

Sajid Bashir[1,2], James Dinn[1], Jingbo Liu[1,2]
1: Texas A&M University-Kingsville, Kingsville, TX 78363, 361-593-2919 (ph), 361-593-3597
(fax), james.dinn@students.tamuk.edu, kfjll00@tamuk.edu; br9@tamuk.edu;
2: Advanced Light Source, Lawrence Berkeley National Laboratory, Berkeley, CA

ABSTRACT

Metallic silver nanoparticles (NPs) have extensively been used in the treatment of disease and purification and heralded the 'first wave' of disinfection science, the 'second wave' being the nanocomposite of metal-doped TiO_2. Recent advances in engineered surfaces have enabled ultrahigh surface area and rapid sterilization via using metal-organic frameworks (MOFs) as the 'third wave' disinfectant. MOFs offer the same advantages as colloids but also have ultra high surface area, long term persistence and ultra low doses, applied for water purification.

INTRODUCTION[§]

Semmelweis [1], Pasteur [2] and Lister [3] had demonstrated that microbes could cause disease and that disinfection through chemical means such as with bleach or heat treatment (pasteurization), greatly reduced the transmission of disease [4]. The approach required for "standards" in *in vitro* "procedures" [5] using specific strains such as *Staphylococcus aurerus* (ATCC 6538), Pseudomonas aeruginosa (ATCC 15422) using the proscribed dilution method [6], although other microorganism were also evaluated such as *Escherichia coli* [7]. These protocols firmly established the profound benefits of sterilization [8] and antisepsis (for living matter) [9]. As early as 1900, a review of the literature in a report to the committee on disinfectants identified the following candidates: Formaldehyde, Mercuric chloride, Chloride of lime, Sulphurous acid, Phenol, Copper sulphate, Zinc chloride, Quick lime and boiling of water [10]. Chemical disinfectants are chemicals, which inhibit and inactivate microbes. Since microbes are living biological organism, which may use oxygen as their terminal electron acceptor (aerobic respiration). Small molecules in the absence of oxygen (anaerobic respiration), possible intervention at physical or electrochemical disruption can lead to inhibition of growth and cellular inactivation. The process is achieved chemically are numerous and are classified based on the functional group to which the chemical disinfectant belongs to, or the area of action leading to inhibition e.g. acid, base, detergent, respiration inhibitor and so forth [11]. In this study, three generations of nanodisinfectants have been developed: core-shelled silver Ag NPs (1st generation); Ag-titania (2nd generation); and nanoscaled MOFs (3rd generation). These nano-materials can be readily prepared using feasible wet-chemistry method [12]. The rationale for using nanomaterials is that their mode of action is different from bleach and may be applicable in circumstances, where microbes have developed a tolerance, particularly when low concentrations of bleach are used for disinfection.

EXPERIMENTAL

A green chemistry method using natural product as reducing agents was employed to produce nanoscaled metal silver (Ag) and Ag-modified titania (TiO_2). The second approach, a hydro-solvothermal, was employed to prepare MOFs as disinfectants. Gram-negative bacteria, *Escherichia coli* (*E. coli*) and Gram-positive bacteria, *Staphylococcus aureus* (*S. aureus*) were selected to determine the antibacterial activities of these three types of anti-bacterial materials. Briefly, the nanodisinfectants were incubated with microbes of a population which gave an optical density (O.D.) absorbance of > 0.6 at 600 nm. Tecnai F20-G^2, equipped with energy loss spectroscopy mode was used to conduct the morphological alteration of bacteria after treatment with nano-disinfectants. The X-ray diffraction studies were also conducted to determine the crystalline structure of nanomaterials.

DISCUSSION
Application of Metal Nanoparticle as Disinfectants

Figure 1 demonstrates that silver core-shelled nanoparticles (NPs) inactivated the *E. coli*, with high potency. The minimum bactericidal concentration (MBC) was used to evaluate the bactericidal potency and found to be less than 40 ppm. This effectiveness is tangentially related to synthetic method, for example silver nanoparticles (with average size of 12.5 nm) reduced with (1:1 molar ratio). Citrate reducing agents were effective at 40 ppm, whereas similar particles reduced with (1:2 molar ratio) citrate or dimethylamine borane were effective at lower concentration (1:2 NP: reductant, 100% inhibition using 10 ppm). The most effective formulation in terms of speed and dosage 1:1 (NP: reductant) at 10 ppm were effective in 2 hr. It was also found longer incubation time corresponded to lower dosage. A progression from 2→4→6→8 hr incubation in terms of dosage in ppm was found to be 40→10→2.5→0.6 to inactivate *E. coli* respectfully ([13]). Lastly, the actual time/dosage is microbe-dependent; comparison to *S. aureus* (2.5 ppm) was the most effective dose, whilst *S. aureus* always required a higher lethal dose relative to *E coli*. BacLight staining also showed that microbial morphology was in the whole intact with silver nanoparticles, whereas with bleach, total cellular degradation was observed. Elemental mapping for potassium (L$_3$ edge at 294 eV), [13] supports our hypothesis that silver aids directly or indirectly the depolarization of the microbial membrane, resulting in K-leakage, which is consistent with older literature and is summarized in **Figure 1.**

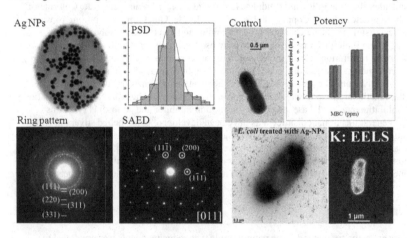

Figure 1. The morphological analyses of bacteria treated using Ag NPs disinfectants; top row: left-to-right: scanning electron microscope (SEM) images of NP showing size similarity; particle size distribution, with a modal diameter of 25 nm; transmission electron microscope (TEM) images of a untreated *E. coli* microbe, as control with intact morphology and flagella and plot of MBC (in ppm of silver, Ag) versus disinfection time (in hr) at four concentrations: 40 ppm (blue column), 10 ppm, (red column), 2.5 ppm (green column) and 0.6 ppm (purple column. Bottom row: left-to-right: Ring pattern of Ag, showing high degree of crystallinity; Selected area electron diffraction (SAED) spectra showed deoxyribonucleic acid (DNA) migration (dark and light regions) and electron energy loss spectroscopy (EELS) map of the *E. coli* surface for potassium. Partial figure adapted from [Chamakura *et al.*, Colloids Surf. B: Biointerfaces, 2011, 84B(1), 1047-1054]- Reproduced by permission of Elsevier, copy-right license #:3094990649632 and [Medina-Ramirez *et al.*, Colloids Surf. B: Biointerfaces, 2009, 73B (2), 185-191]- Reproduced by permission of Elsevier, copy-right license #:3094990866029.

Application of Nanocomposite Disinfectants

A comparison of disinfection effectiveness between core-shelled Ag, Ag/TiO$_2$ NPs, and bleach has revealed a number of facts. Over a 2-week period, the MBCs were determined to be 2.5 ppm for Ag (**Figure 1**) and 0.6 ppm for Ag/TiO$_2$ for 100% inhibition of Gram-negative and Gram-positive. One drawback of NPs is that the time needed to disinfect infectious agents is ~ 2 hours at ~10 ppm and 6 hours at < 1 ppm (**Figure 2**). A major difference between core-shelled Ag and Ag/TiO$_2$ is that Ag/TiO$_2$ is cheaper than Ag and is more effective under visible light conditions. By comparison, bleach acts rapidly and at a minimum dose of 50 ppth for up to 3 days in sunlight [14]. At 50 ppth of bleach, disinfection is achieved through total microbial disintegration, whereas Ag and Ag/TiO$_2$ retain microbe morphology to some degree. The NP strategy affords two distinct advantages: Much lower concentration of Ag and Ag/TiO$_2$ was needed for disinfection (ppm vs. ppth); and prolonged environmental persistence was achieved (weeks vs days) compared to bleach.

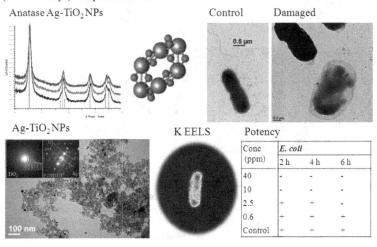

Anatase Ag-TiO$_2$ NPs   Control   Damaged

Ag-TiO$_2$ NPs   K EELS   Potency

Conc (ppm)	*E. coli*		
	2 h	4 h	6 h
40	-	-	-
10	-	-	-
2.5	+	+	-
0.6	+	+	+
Control	+	+	+

Figure 2. Top-left X-ray diffraction spectra of Ag-TiO$_2$ with unit structure model; TEM image of untreated *E. coli*, as control (center) and treated with NPs showing DNA migration and cell wall peeling (right). Bottom-left: An TEM of the nanocomposite, with ring pattern and SAED (inserts), EELS mapping for K (center) and a table of concentration (in ppm) versus incubation time (in h) for *E. coli* with Ag-NPs, showing whether 100% inactivation was achieved (-) or not achieved (+). [Partial figure or table adapted from [Liu *et al.*, Biomater. Sci., 2013,1, 194-201] - Reproduced by permission of The Royal Society of Chemistry, [Medina-Ramirez *et al.*, Dalton Trans., 2011,40, 1047-1054] - Reproduced by permission of The Royal Society of Chemistry and [Chamakura *et al.*, Colloids Surf. B: Biointerfaces, 2011, 84B(1), 1047-1054]- Reproduced by permission of Elsevier, copy-right license #:3094990649632.

Application of Metal Organic-Framework (MOF)Disinfectants[‡]

MOFs are crystalline compounds that consist of metal ions coordinated with organic linkers to form structured frameworks with specific properties. MOFs are designed to oxidize and depolarize outer microbial membranes through the central atom, further to inhibit protein synthesis. Using smart-design, MOFs can be generated to target specific components involved in water purification, such as the removal of pathogens, organics, and metal ions [15]. Due to their ultrahigh internal surface area, small molecules can be encapsulated and simply released by a change in environmental pH. Therefore, optimal effectiveness can be chemically engineered into MOFs, which can be used at low concentrations for extended duration (**Figure 3**) representing the "third-wave" of disinfectant.

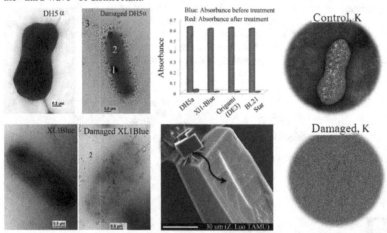

Figure 3. Top-row: left-to-right: SEM Image of untreated *E. coli* as control; *E. coli* treated with MOFs showing lysis, a plot of four different strains of *E. coli* with controls, showing that the MOFs was able to inactivate them all, at a similar concentration; and EELS K map of a control *E. coli* showing no leakage. Low-bottom: left-to-right: A control *E. coli* showing no damage; the same strain with a MOF, showing extensive damage, a SEM image of a MOF crystal face (black arrow) with a MOF cluster (insert); and EELS K map of damage cell, showing extensive

leakage, since cell surface and background appear to be identical. Partial figure or table adapted from [‡ Zhuang *et al.*, Adv. Healthcare Mater., 2012,1(2), 225-238] - Reproduced by permission of Wiley, copy-right license #:3095000312659 and from [§ Liu *et al.*, Historical Overview of the First Two Waves of Bactericidal Agents and Development of the Third Wave of Potent Disinfectants, Chapter 6, pp 129–154 in Nanomaterials for Biomedicine, ACS Symposium Series, Vol. 1119, Ed. R. Nagarajan]. Copyright (2012) American Chemical Society.

Potency Comparison of Various Disinfectants

Nanoparticles concentration for 100% microbial inhibition is microbe sensitive. Microbes, which possess a thicker cell wall, require higher concentrations than thinner cell wall microbes, which are not observed for either phenol or bleach, indicating that mode if inactivation may be different between NP-induced disinfection and bleach/phenol induced toxicity. The dose also appears to be independent on the incubation-time for the chemicals (**Figure 4**), which is not the case for NPs. The longer the incubation time, the lower the lethal NP dose was obtained. Elemental mapping for potassium (L_3 edge at 294 eV) and other heteroatom's from the main and transition block support the hypothesis that one mode of bacterial inactivation is through membrane depolarization, which we believe is through metal induced redox. A comparison of bleach, silver, silver-titania and MOF indicate similarities and differences. Similarity within the nanoparticles/MOF disinfectants are that they require much lower doses (lower ppms) to be 100% lethal, whereas bleach required a much higher lethal dose. Similarity between MOFs and bleach are there rapid inactivation times compared to NPs. Overall, MOF exhibits the advantages of low dose, long persistence (like NPs) and rapid inactivation (like bleach) without the drawbacks of disinfection by-products.

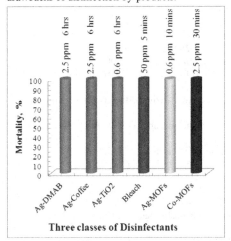

Figure 4. A comparison of the lethal dose for 100% inhibition of *E. coli* with different formulations of silver using different reducing agent (labeled as "Ag-reductant"), a single formulation of silver-titania and two different metals in MOFs. Partial figure or table adapted

CONCLUSIONS

We demonstrate a new paradigm in microbial inhibition as a tiered-strategy in water purification. The first two generations represent technological advances that allow for inexpensive, efficient water disinfection. MOFs represent the 3rd generation, yielding the same advantages as nanomaterials but exhibit enhanced and more rapid antimicrobial activity. Importantly, these nanomaterials can serve as model systems to develop advanced water disinfection technologies. These technologies can be used to address the needs relevant to the 'Water for Life' decade to improve human access to portable clean water at reduced cost [15].

ACKNOWLEDGMENTS

The authors wish to thank the College of Arts & Sciences (160336-00002) and Welch Departmental Grant (AC-0006) at Texas A&M University-Kingsville (TAMUK) for funding and student support respectfully. The Microscopy and Imaging Center (MIC) at TAMU and the Department of Chemistry at TAMUK are also duly acknowledged for their technical support and nanostructure characterization.

REFERENCES

1. H. Wykilicy, M., *Infection Control* **4**, 367 (1983).
2. P. Debre Industrial Pasteurization, In Louis Pasteur, (Ed. P. Debre'), The John Hopkins University Press, Baltimore, MD, USA, (1988).
3. D:G. M. Lister, *the New England Journal of Medicine* **294**, 1286 (1976).
4. S. Ridel, and J.T.A. Walker, *Journal Royal Sanitary Institute* **24**, 424 (1903).
5. Association of Official Analytical Chemists. 1990. Official Methods of Analysis, 15th ed. AOAC, Arlington, VA, (also see
   http://www.fda.gov/Food/ScienceResearch/LaboratoryMethods/BacteriologicalAnalytical ManualBAM/ucm063346.htm).
6. E.C. Cole, W.A. Rutala, L. Nessen, N.S Wannamaker, and D.J Weber, *Applied and Environmental Microbiology* **56**, 1813 (1990).
7. F. Heimets, W.W. Taylor, and J.J. Lehman, *Journal of Bacteriology* **67**, 5, (1954).
8. M.R. Holland, *Oral Surgery, Oral Medicine, Oral Pathology* **8**, 788 (1955).
9. E.L. Larson, *American Journal of Infection Control* **23**, 251 (1995).
10. F.C. Robinson, *Public Health Papers and Reports* **26**, 151 (1900).
11. R.J. Romani, *Advances in Food Research* **15**, 57 (1966).
12. W Zhuang, D Yuan, J.-R Li, Z Luo, H.-C. Zhou, and S Bashir, and J Liu, *Advanced Healthcare Materials* **1**, 225, (2012).
13. K Chamakura, R Perez-Ballestero, Z Luo, S Bashir, and J Liu, *Journal of Colloids Surface B: Biointerfaces* **84**, 88 (2011).
14. W.A. Rutala, and D.J. Weber, *Clinical Microbiology Reviews,* **10**, 597 (1997).
15. R. Rheingans, R. Dreibelbis, and M.C. Freeman, *Global Public Health* **1**, 31 (2006).

Mater. Res. Soc. Symp. Proc. Vol. 1498 © 2012 Materials Research Society
DOI: 10.1557/opl.2012.1652

Anodic Aluminum Oxide (AAO) Membranes for Neurite Outgrowth

Meghan E. Casey[1], Anthony P. Ventura[2,3], Wojciech Z. Misiolek[2,3] and Sabrina Jedlicka[1,2,4]

[1]Bioengineering, [2]Materials Science and Engineering, [3]Institute for Metal Forming, [4]Center for Advanced Materials and Nanotechnology
Lehigh University, Bethlehem, PA 18015 USA

ABSTRACT

Anodic aluminum oxide (AAO) membranes were fabricated in a mild two-step anodization procedure. The voltage was varied during both anodization steps to control the pore size and morphology of the AAO membranes. Pore sizes ranged from 34 nm to 117 nm. Characterization of the pore structure was performed by scanning electron microscopy (SEM). To assess the potential of the AAO membranes as a neuronal differentiation platform, C17.2 neural stem cells (NSCs), an immortalized and multipotent cell line, were used. The NSCs were forced to differentiate via serum-withdrawal. Cellular growth was characterized by immunocytochemistry (ICC) and SEM. ImageJ software was used to obtain phenotypic cell counts and neurite outgrowth lengths. Results indicate a highly tunable correlation between AAO nanopore sizes and differentiated cell populations. By selecting AAO membranes with specific pore size ranges, control of neuronal network density and neurite outgrowth length was achieved.

INTRODUCTION

The ability to generate and control neuronal growth *in vitro* from stem cell precursors is a growing area of interest in biomedicine. Neuronal development occurs in various stages from immature precursor cells to fully integrated and functionally mature neurons [1]. These developmental steps are classified into two categories: activity independent and activity dependent. Independent landmarks are thought to be genetically determined and include neuronal differentiation, migration and axon guidance [2]. Activity dependent stages of neuronal growth are heavily regulated by secreted molecules such as hormones and neurotransmitters [2]. The overall effects of the secreted molecules *in vivo* are well researched; however, the effects on *in vitro* differentiation are not fully understood. Researchers are unable to identify and measure the small molecules *in vitro*, as the secreted hormones are absorbed by neighboring cells. Therefore, a substantial opportunity exists in neuronal interface research to develop a material platform that allows for both the proliferation and differentiation of stem cells into neurons and the ability to quantify the secretome of neuronal cells.

C17.2 neural stem cells (NSCs) are an immortalized and multipotent cell line established by Snyder *et al.* [3,4]. Derived from the external germinal layer of neonatal mouse cerebellum, C17.2 neural precursors have been show to successfully implant into mouse germinal zones [4]. The NSCs integrate into the implanted tissue and contribute to cerebellum development [4].

Because C17.2 cells are functional *in vivo*, they are a relevant cell model for studying neuronal platforms.

AAO membranes are biocompatible and composed of highly-ordered nanopores that penetrate the entire material [5-7]. The inert properties of AAO membranes support the growth of neuronal cells and the nanopores may allow for selective concentration of secreted molecules. Nanopore sizing, surface functionalization and morphologies are controllable based upon experimental parameters and allow for precise segregation and selection of secreted molecules [8]. In this work, neuronal differentiation of C17.2 neural stem cells on AAO membranes is examined as a means for developing an artificial cell/material synapse system.

EXPERIMENTAL DETAILS
Anodic Aluminum Oxide Membrane Fabrication
The fabrication procedure for the AAO membranes used in this experiment was based on a two-step mild anodization procedure originally proposed by Masuda and Fukuda in 1995 [9]. 99.99% pure aluminum was electropolished at a current of 6.5 A to remove any surface scratches. The electrolyte used for both anodization steps was 2.7% oxalic acid by weight mixed with ethanol in a ratio of 5:1. The temperature was held constant at 0°C during anodization with a recirculating chiller and air stirring.

The first anodization step was conducted for a total of 2 hours to form ordered pore nucleation sites. The oxide layer was then etched away at 65°C for 1 hour in a 1:1 mixture of 8% by volume orthophosphoric acid and 4% by weight chromic acid. The second anodization step was carried out for approximately 35 hours. The second anodization has a much more stable growth pattern because of the already established nucleation sites leading to an ordered pore structure, but there is no significant benefit to further etch-anodization cycles in terms of pore structure [10]. The voltage was varied during the first and second anodization steps in order to control the pore size and morphology of the AAO membranes as proposed by Bai *et al.* [7]. An etch cycle in a 1:1 mixture of 10% hydrochloric acid by volume and 0.1 M CuCl removed the remaining aluminum on the back of the foil leaving only the AAO membrane. The final etching step in 0.1 M orthophosphoric acid at 30°C for 75 minutes removed the barrier layer resulting in membranes with straight channel through-pores for cell growth. Membranes with degraded pore structures (pignosed) were also prepared for cell growth to observe the effect of pore ordering on neuron development. Resulting pore structures were quantified using SEM and ImageJ software.

Materials Sterilization and Preparation
The AAO membranes were UV sterilized overnight. The membranes were washed with sterile 1X Phosphate Buffered Saline (PBS), followed by a 30 minute wash with growth medium (GM) [high glucose Dulbecco's Modified Eagle Medium (DMEM), supplemented with 10% Fetal Bovine Serum (FBS), 5% Horse Serum (HS), 1% L-glutamine]. The membranes were incubated in GM overnight at 37°C and 5% CO_2.

Cell Culture

C17.2 NSCs were routinely maintained in GM. The NSCs were seeded onto the AAO membranes at 10,000 cells/cm^2 and incubated at 37°C and 5% CO_2. The cells grew on the membranes for 2 days before starting the serum-withdrawal protocol to differentiate the NSCs. Half of the total volume of media was removed and replaced with serum-free (DMEM high glucose with 1% L-glutamine) media every 2 days. The cells grew on the membranes for 14 days after the serum concentration had dropped below 1%.

Immunocytochemistry (ICC)

Following the serum withdrawal, the cells were formalin fixed and analyzed for neuronal and astrocytic differentiation using β-tubulin III (1:1000, Covance #A488-435L, neuronal) and glial fibrillary acidic protein (GFAP) (1:500, Sigma #C9205, astrocytic) antibodies, respectively. Additional samples were analyzed with nestin (1:100, Developmental Studies Hybridoma Bank #Rat-401, NSCs) and neurofilament H/M (1:100, Covance #SMI-33R, neuronal) antibodies. Nuclei were counter-stained with Hoechst 33258.

SEM Preparation

After the cells grew on the AAO membranes, the samples were prepared for SEM. The samples were fixed in 5% glutaraldehyde. Following fixation, the samples underwent dehydration via ethanol and hexamethyldisilazane incubations. After air drying overnight, the samples were sputter coated with iridium to provide a conductive surface for SEM observation.

RESULTS AND DISCUSSION

The NSCs that underwent the differentiation process were examined using ICC and SEM. All samples were positive for β-tubulin III, nestin and neurofilament H/M, indicating a mixed phenotype population (Figure 1); GFAP did not result in positive staining, indicating no astrocytic differentiation.

Figure 1. Differentiated C17.2s on AAO membranes stained for expression of nuclei (blue, circles). Expression of (a) neurofilament H/M and (b) nestin are shown in red (extensions).

The β-tubulin III stained samples were used for neurite measurements in NeuronJ (Figure 2), as neurite outgrowth is often correlated to increased neuronal differentiation [11]. Compared to the tissue culture treated glass controls, all AAO membranes supported enhanced neurite outgrowth. The samples with the longest measured outgrowths were within the 64-78 nm range, suggesting the possibility of an optimal range of pore sizes (Figure 3).

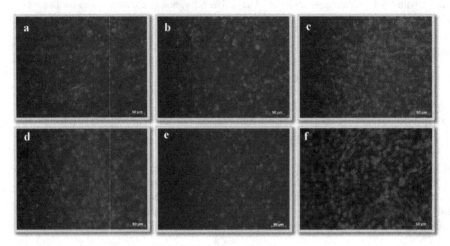

Figure 2. C17.2s differentiated on AAO membranes and stained for β-tubulin III (green, extensions) and nuclei (blue, circles). AAO membrane pore sizes are (a) 38 nm, (b) 64 nm, (c) 77 nm (pignosed), (d) 78 nm and (e) 117 nm (pignosed). Tissue cultured treated glass control is shown in (f).

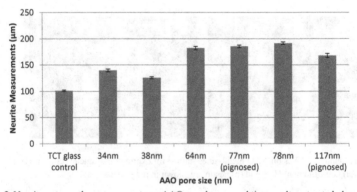

Figure 3. Neurite outgrowth measurements on AAO membranes and tissue culture treated glass (TCT).

Samples were also analyzed for neuronal population percentages (Figure 4). Tissue culture treated glass controls resulted in the highest neuronal population. Additionally, AAO membranes with smaller pore sizes (34 and 38 nm) supported larger neuronal populations than the other ranges of pores (64-78 nm and 117 nm). These results suggest a relationship between pore size and neuronal population dynamics.

Cell layer morphology was examined via SEM. Figures 4a and 4b illustrate dense, tissue-like cell growth on AAO membranes. The tissue layer was mixed in phenotype, as observable through morphological differences in the images. Neurons are outstretched and interacting with both the underlying cell layer and the AAO membrane (Figures 4c, 4d, 4e).

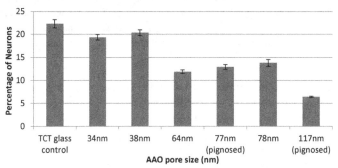

Figure 4. Neuronal population percentage of C17.2 cells differentiated on AAO membranes and TCT.

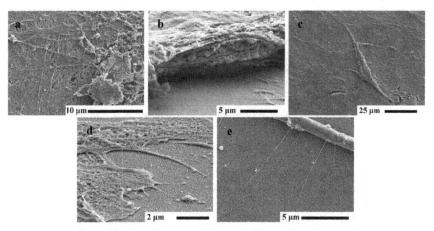

Figure 5. SEM images of differentiated C17.2 cells grown on AAO membranes. Images illustrate dense cellular growth (a, b), with neuronal interactions on underlying cell layers (c) and AAO membranes (d, e).

CONCLUSIONS

The resulting ICC images indicate that neurite outgrowths were the greatest on AAO membranes with pore sizes of 64, 77 and 78 nm. Outgrowths on 38 and 117 nm were greater than those measured on tissue culture treated glass. Neuronal percentage populations also changed based upon the membrane pore size. As illustrated by ICC and SEM images, the resulting cell layer is dense, with a mixed phenotype population.

The resulting data suggest that AAO membranes support greater neurite outgrowth than traditional cell culture surfaces such as tissue culture treated glass. The presence of the dense, mixed phenotype population suggests the possibility of tissue growth on the membranes. The combinatorial results indicate the AAO membrane pore size can directly affect the differentiated cell population. Data suggests a highly tunable correlation between AAO nanopore sizes and differentiated cell populations. By selecting AAO membranes with specific nanopore size ranges, control of neuronal network density and neurite outgrowth length was achieved.

ACKNOWLEDGEMENTS

The research was supported by NSF CBET #1014957 and the Lehigh University Faculty Innovation Grant 2011. We also thank Evan Snyder of the Sanford-Burnham Medical Research Institute for providing the C17.2 neural stem cells.

REFERENCES

1. G. Kempermann, S. Jessberger, B. Steiner, G. Kronenberg, *Trends Neurosci.* **27**, 447-452 (2004).
2. D. C. Lie, H. Song, S. A. Colamarino, G. Ming and F. H. Gage, *Annu. Rev. Pharmacol. Toxicol.* **44**, 399-421 (2004).
3. E. Y. Snyder, D. L. Deitcher, C. Walsh, S. Arnold-Aldea, E. A. Hartwieg and C. L. Cepko, *Cell* **68**, 33-51 (1992).
4. E. Y. Snyder, C. Yoon, J. D. Flax and J. D. Macklis, *Proc. Natl. Acad. Sci.* **94**, 11663-11668 (1997).
5. F. Li, L. Zhang and R. M. Metzger, *Chem. Mater.* **10**, 2470-2480 (1998).
6. A. Belwalkar, E. Grasing, W. Van Geertruyden, Z. Huang and W. Z. Misiolek, *J Memb Sci.* **319**, 192-198 (2008).
7. A. Bai, C. Hu, Y. Yang and C. Lin, *Electrochimica Acta* **53**, 2258-2264 (2008).
8. J. Hu, J. H. Tian, J. Shi, F. Zhang. D. L. He, L. Liu, D. J. Jung, J. B. Bai and Y. Chen, *Microelectronic Engineering* **88**, 1714-1717 (2011).
9. H. Masuda and K. Fukuda, *Science* **268**, 1466-1468 (1995).
10. G. D. Sulka, S. Stroobants, V. Moshchalkov, G. Borghs and J. P. Celis, *J Electrochem. Soc.* **149**, D97-D103 (2002).
11. J. Jaworski, S. Spangler, D. P. Seeburg, C. C. Hoogenraad and M. Sheng, *J Neurosci.* **25**, 11300-11312 (2005).

Mater. Res. Soc. Symp. Proc. Vol. 1498 © 2013 Materials Research Society
DOI: 10.1557/opl.2013.12

Investigation of biocompatibility on nitrogen-doped a-C:H film coating scaffold surface in *in-vivo* and *in-vitro* tests

Yasuharu Ohgoe[1], Tomoaki Wada[1], Yasuyuki Shiraishi[2], Hidekazu Miura[2], Kenji K. Hirakuri[3], Akio Funakubo[1], Tomoyuki Yambe[2], and Yasuhiro Fukui[1]
[1] Division of Electronic and Mechanical Engineering, Tokyo Denki University,
Ishizaka, Hatoyama, Saitama, 350-0394 Japan
[2] Institute of Development, Aging and Cancer, Tohoku University
Seiryo-machi 4-1, Aoba-ku Sendai, 980-8575 Japan
[3] Department of Electrical Engineering, Tokyo Denki University,
Senju Asahi-cho 5, Adachi-ku, Tokyo 120-8551 Japan

ABSTRACT

In this study, in order to investigate biocompatibility of nitrogen-doped hydrogenated amorphous carbon (a-C:H:N) film coating segmented polyurethane (SPU) scaffold fiber sheet (a-C:H:N-Scaffold) in *in-vitro* test, mouse fibroblasts (NIH 3T3) cells were grown on the a-C:H:N-Scaffold. The cell behavior was monitored by time-lapse imaging system. Additionally, the a-C:H:N-Scaffold was implanted at partial aorta descendens of a goat for 35 days. The surface morphology, composition, and wettability of the a-C:H:N-scaffold was estimated by Scanning Electron Microscope (SEM), X-ray photoelectron spectrometer (XPS), and contact angle measurement. In *in-vitro* test, it was observed that a-C:H:N film coating had a facilitatory effect on cell motility and cell growth. In *in-vivo* test, it was observed that the a-C:H:N-Scaffold surface was uniformly covered by neointima. The a-C:H:N-Scaffold surface had no thrombus formation as an inflammatory reaction and it was shown that the a-C:H:N film coating had a good blood compatibility. These results suggest that a-C:H:N film coating has good cytocompatibility and blood compatibility and it is a promising approach for improvement of biocompatibility of biomaterial surfaces.

INTRODUCTION

Demand for vascular grafts in modern medicine and surgery is vast and huge, and synthetic vascular grafts should have biocompatibility, anti-thrombosis, anti-infective, and living body without toxicity and/or carcinogenicity, etc [1, 2]. However, in some cases, synthetic vascular grafts are lack of cell growth for biocompatibility [1, 2]. The lack of biocompatibility causes substantial risk of infection or thrombosis problems during long implantation periods. In order to minimize or avoid the risks, tissue engineering, whereby living tissue replacements can be constructed, has emerged as a solution to some of the problems [3].

Fibrous scaffolds are key components in tissue engineering and typically fabricated in the form of a biological matrix or material and have been used for a variety of biomedical applications such as vascular graft, bone, ligament, skin, neural tissues, and skeletal muscle, etc. and as vehicle for the controlled delivery of drugs, proteins, and DNA [4]. Typical scaffolds which are made of polymeric biomaterial (non-biodegradable or biodegradable) provide the structural support for cell attachment and subsequent tissue development [4, 5]. Moreover, the scaffolds are defined as three-dimension porous solid biomaterials designed to perform the following functions; promote cell-biomaterial interactions, cell adhesion, proliferation, and inhibition of inflammation or toxicity *in-vivo* [4]. However, most of the fibrous scaffolds do not possess any specific functional groups, and they must be optimized surface conditions for successful applications.

In our previous works, we have developed hydrogenated amorphous carbon (a-C:H) film deposition technique for cell growth and deposited on micro segmented polyurethane (SPU) scaffold fibers [6]. We investigated structural and compositional effects of the a-C:H film on cell growth. It was observed that the a-C:H film coating enhanced cell growth. As well known, application of carbon coatings for medical engineering has indicated that it is promising for improvement of biomaterials [7-12]. Especially, a-C:H film coating is one of materials of highest biocompatibility and often used as medical devices. There are many reports suggested available applications of a-C:H films for medical devices. a-C:H film including diamond-like carbon (DLC) has low frictional coefficient, high wear and corrosion resistance, chemical inertness, high electrical resistivity, infrared-transparency, high refractive index, and excellent smoothness, etc. [7-12]. Additionally, a-C:H film coating has emerged as a potential technique for blood interfacing applications [7-9]. Several recent attempts were made to modify the characteristics of a-C:H coating by adding elements, such as O, N, and F, into the films [13, 14]. Especially, nitrogen-doped a-C:H (a-C:H:N) film coating has indicated promise for hasten the cell growth applications [14]. a-C:H:N film is expected to have good cellular responses since are highly biocompatible.

In this study, in order to investigate an effect of nitrogen dope for a-C:H film on cell growth, we observed cell behavior on the a-C:H:N film coating SPU micro fibers scaffold, based on the cell cultures (*in-vitro test*). Moreover, we investigated biocompatibility of the a-C:H:N-Scaffold in *in-vivo* test.

EXPERIMENT

Fabrication of SUP scaffold fiber and a-C:H:N film coating

In this study, SPU micro fiber scaffold was fabricated by electronspinning method. The SUP (Mn = 80,000, Mw = 160,000, NKY-26, UBE Industries, LTD) dissolved in mixture solution of Tetrahydrofuran / N N-dimethyl formamide (8:2). The SPU concentration of the solution was 14.5 wt% and was drawn into fibers by the high voltage (12.5 kV), and SPU scaffold with a fiber diameter of 1.5 μm was fabricated.

After the preparation of scaffold sample, a-C:H:N film was deposited on the glass substrate with the scaffold fiber by radio frequency (r.f.) plasma chemical vapor deposition (CVD) technique. Schematic diagram is shown in figure 1. The r.f. power was kept at 100 W constantly, and decomposed mixture gas of CH_4/N_2 gas (4:6) at 100 Pa with a deposition time of 5 minutes constantly (a-C:H:N-Scaffold). On the other hand, a-C:H film was deposited as same as the a-C:H:N film deposition conditions without N_2 gas (CH_4 gas only, a-C:H-Scaffold). The surface morphology and compositions of the a-C:H:N-Scaffold, a-C:H-Scaffold, and Scaffold without coating were estimated by scanning electron microscope (SEM) image and X-ray photoelectron spectrometer with Mg $K\alpha$ radiation (XPS), respectively. Additionally, wettability of each scaffold sheet which had 500 μm thicknesses was evaluated by contact angle measurements. The water contact angle was measured by CCD camera with demineralized water where the camera can catch the contact angle in left and right side (2-μL water droplets, temperature: 20 °C, relative humidity: 20%), respectively.

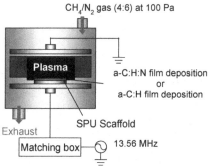

CH₄/N₂ gas (4:6) at 100 Pa

Plasma

a-C:H:N film deposition
or
a-C:H film deposition

SPU Scaffold

Exhaust

Matching box — 13.56 MHz

Figure 1 a-C:H:N and a-C:H film deposition on SPU scaffold by r.f. plasma CVD method

In-vitro test

The SPU scaffold was fabricated on a glass substrate (φ200 mm), and the glass substrate was put on a polystyrene (PS) dish for cell culture. a-C:H:N and a-C:H film was deposited on the scaffold fiber together with the glass substrate on the PS dish, respectively. Fabrication of the scaffold samples and a-C:H:N and a-C:H film coatings were carried out under the above methods. NIH-3T3 cells were used to estimate the cellular responses on the a-C:H:N-Scaffold, a-C:H-Scaffold, and Scaffold without coating. The cells were seeded on these scaffold sample surfaces, prepared in D-MEM solution, and adjusted to a density of 0.5×10^3 cell/cm². The cell culture was incubated at 37 °C in an atmosphere consisting of 5% CO_2 and 95% air with a relative humidity of 100%. The cells were cultured in the D-MEM for 3 days. During the cell culture, cell behavior on the scaffold fiber was analyzed at intervals of 10 minutes for 72 hours by time-lapse imaging system.

In-vivo test

Hemocompatibility of a synthetic vascular graft is related to materials and components. To improve hemocompatibility of a synthetic vascular graft, not only design of structure but also blood contacting surface must be improved. The purpose of this *in-vivo* test was to evaluate hemocompatibility of a-C:H:N-Scaffold, a-C:H-Scaffold, and Scaffold without coating. a-C:H:N and a-C:H film was deposited on the scaffold sheet, respectively. The thickness and diameter of the scaffold sheets were 500 μm and φ10 mm, respectively. The fabrication of the scaffold sheets and a-C:H:N and a-C:H film coating were carried out under the above methods. And then, these scaffold sheets were implanted at partial aorta descendens of goat for 35 days.

RESULTS AND DISCUSSION

Fabrication of SUP scaffold fiber and a-C:H:N film coating

Figure 2 shows the surface morphology of a-C:H:N-Scaffold, a-C:H-Scaffold, and Scaffold without coating. Each fiber surface of the scaffolds was rough after the a-C:H:N film or a-C:H film coating uniformly. There was no significant difference between a-C:H:N and a-C:H coating on the scaffold surface. In the XPS analysis, carbon (C_{1s}), oxygen (O_{1s}), and nitrogen (N_{1s}) spectra for a-C:H:N-Scaffold, a-C:H-Scaffold, and Scaffold sheet were observed (Figure 3).

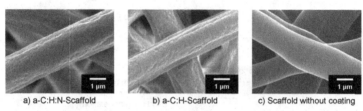

a) a-C:H:N-Scaffold   b) a-C:H-Scaffold   c) Scaffold without coating

Figure 2 Surface morphology of a-C:H:N-Scaffold, a-C:H-Scaffold, and Scaffold without coating fiber (SEM images).

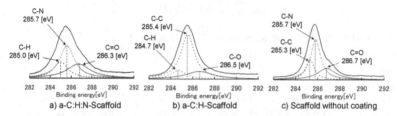

a) a-C:H:N-Scaffold   b) a-C:H-Scaffold   c) Scaffold without coating

Figure 3 Chemical characterization of a-C:H:N-Scaffold, a-C:H-Scaffold, and Scaffold without coating surfaces by XPS (C_{1s}) spectra.

In comparison with a-C:H:N and a-C:H coating, the nitrogen-dope replaced C-O by C=O bounds. Moreover, in investigation of wettability, the surface of the a-C:H:N-Scaffold was changed to hydrophilic side by the nitrogen-dope. The contact angle of a-C:H:N-Scaffold, a-C:H-Scaffold, and Scaffold without coating was 106.9°, 123.0°, and 121.4°, respectively. As C=O bonds are known to be hydrophilic [15], the cause is considered to be that the hydrophobic of the a-C:H film surface decreases with increasing in C=O bonds.

In-vitro test

NIH 3T3 cells were cultured on the a-C:H:N-Scaffold in the D-MEM solution. Figure 4 shows the cell behavior image on the scaffold fiber during the cell culture. According to the analysis, it was observed that a-C:H:N film coating had a facilitatory effect on cell motility. The average moving velocity of the cells on the a-C:H:N-Scaffold was higher than other scaffold samples (a-C:H coting and without coating). Additionally, in the cell culture, it was observed that the a-C:H:N-Scaffold had a higher level of cell growth than other scaffold samples as same as moving velocity of the cells.

Majumdar *et al.* presented that increased amount of nitrogen in amorphous hydrogenated carbon nitride (a-H-CN$_x$) film induced an accelerated cell death. Moreover, the authors mentioned surface roughness of the a-H-CN$_x$ film decreases with increase amount of nitrogen in the film and cells did not adhere to the film surface [16]. In cell culture, it is well known that surface roughness and hydrophilic surface is important factor and these have influence on cell adhesion and proliferation, respectively [17, 18]. There is an optimum surface roughness range (20 - 100 nm) that promotes cell adhesion and longevity [19, 20]. In Majumdar's experiments, the a-H-CN$_x$ film had a very small surface roughness on glass substrate. Therefore, it is conceivable that the main factor of the inhibition of cell adhesion was too small surface

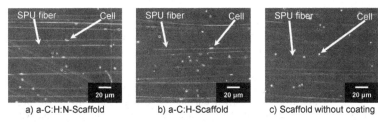

a) a-C:H:N-Scaffold   b) a-C:H-Scaffold   c) Scaffold without coating

Figure 4 Analysis of cell behavior on scaffold fiber.

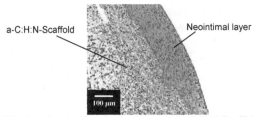

Figure 5 Histology image of neoimtimal layer on a-C:H:N-Scaffold surface.

roughness rather than surface potential. As shown in figure 2, the smooth surface fiber of the SPU scaffold had become noticeably wrinkled by the a-C:H:N and a-C:H film coating. Additionally, a-C:H:N-Scaffold surface had been changed to hydrophilic side by C=O bonds.

The effect of the nitrogen-dope was improvement of physical and chemical characteristics of a-C:H film coating for cell culture on the scaffold. It was observed that a-C:H:N film coating had a facilitatory effect on cell motility and cell growth.

In-vivo test

After the *in-vivo* test, a-C:H:N-Scaffold, a-C:H-Scaffold, and Scaffold without coating surfaces were removed and prepared for paraffin embedding. The blood surface interaction was investigated by hematoxylin-eosin (HE) stain method. Different materials exhibited different patterns of blood-surface interaction. However, in this experiment, each sample surface was absent completely with neointimal coverage. There was no significant difference between a-C:H:N-Scaffold, a-C:H-Scaffold, and Scaffold without coating samples. Histology image of neoimtimal layer on the a-C:H:N-Scaffold surface was shown in figure 5.

For blood compatibility, blood-contacting devices require that stable thrombus formation without inflammatory reaction, and neointimal formation of thrombus layer at blood contacting surface. In figure2, the neointimal layer was organized completely from the aorta side along a layer of the thrombus on each scaffold sheet surface. According to this result, it is expected that a-C:H:N and a-C:H-Scaffold surfaces had a good blood compatibility.

CONCLUSIONS

The a-C:H:N-Scaffold surface had no thrombus formation as an inflammatory reaction and it was shown that the a-C:H:N film coating had a good blood compatibility. Additionally, in *in-vitro* test, it was observed that amount of nitrogen content of the a-C:H:N film was possible to control cells behavior on the scaffold fiber and induced cell growth. These results suggest that a-C:H:N film coating has good cytocompatibility and blood compatibility and it is a promising approach for improvement of biocompatibility of synthetic vascular graft surfaces.

ACKNOWLEDGMENTS

This work was partly supported financially by a Frontier R&D Center of Tokyo Denki University, and part of this work was carried out under the Cooperative Research Project Program of the Institute of Development, Aging and Cancer, Tohoku University, and JGC-S Scholarship Foundation.

REFERENCES

1. C. K. Prasad, and L. K. Krishnan, *Acta Biomaterialia* **4**, 182 (2008).
2. K. K. Johnson, P. D. Russ, J. H. Bair, and G. D. Friend, *AJR* **154**, 405 (1990).
3. C. S. Bahney, C. W. Hsu, J. U. Yoo, J. L. West, and B. Johnstone, *The FASEB Journal* **25**, 1488 (2011).
4. B. P. Chan and K. W. Leon, *European Spine Journal* **17**, 467 (2008).
5. P. C. Caracciolo, A. A. A. de Queiroz, O. Z. Higa, F. Buffa, G. A. Abraham, *Acta Biomaterialia* **4**, 976 (2008).
6. Y. Ohgoe, H. Matsuo, K. Nonaka, T. Yaguchi, K. Kanasugi, K. K. Hirakuri, A. Funakubo, and Y. Fukui, *Mater. Res. Soc. Symp. Proc.* **1138**, 1138-FF12-03 (2011).
7. J. M. Lackner, W. Waldhauser, *BHM* **155**, 528 (2010).
8. H. C. Cheng, S. Y. Chiou, C. M. Liu, M. H. Lin, C. C. Chen, K. L. Ou, *J. Alloys Compd.* **477**, 931 (2009)
9. R. K. Roy, Sk. F. Ahmed, J. W. Yi, M.-W. Moon, K.-R. Lee, Y. Jun, *Vacuum* **83**, 1179 (2009).
10. W. J. Ma, A. J. Ruys, R. S. Mason, P. J. Martin, A. Bendavid, Z. Liu, M. Ionescu, H. Zreiqat, *Biomaterials* **28**, 1620 (2007).
11. R. K. Roy, K. R. Lee, *J. Biomed. Mater. Res. Part B* **83B**, 72 (2007).
12. J. Robertson, *Diamond Relat. Mater.* **12**, 79 (2003).
13. T. Hasebe, A. Hotta, H Kodama, A. Kamijo, K. Takahashi, T. Suzuki, *New Diamond and Frontier Carbon Technology* **17**, 263 (2007).
14. P. Yang, N. Huang, Y. X. Leng, Z. Q. Yao, H. F. Zhou, M. Maitz, Y. Leng, P. K. Chu, *NIM B* **242**, 22 (2006)
15. Y. Nitta, K. Okamoto, T. Nakatani, H. Hoshi, A. Homma, E. Tatsumi, Y. Taenaka, *Diamond Relat. Mater.* **17**, 1972 (2008).
16. A. Majumdar, K. Schröder, and R. Hippler, *J. Appl. Phys.* **104**, 074702 (2008).
17. O. Kubová, V. Švorčík, J. Heitz, S. Moritz, Ch. Romanin, P. Matějka, A. Macková, *Thin solid films* **515**, 6765 (2007).
18. M. Arnold, E.A. Cavalcanti-Adam, R. Glass, J. Blummel, W. Eck, M. Kantlehner, H. Kessler and J.P. Spatz, *ChemPhysChem* **5**, 383 (2004).
19. E. Biazar, M. Heidari, A. Asefnezhad, N. Montazeri, *Int. J. Nanomedicine* **6**, 631-639 (2011).
20. S. P. Khan, G. G. Auner, G. M. Newaz, *Nanomedicine:NBM* **1**, 135-129 (2005).

Mater. Res. Soc. Symp. Proc. Vol. 1498 © 2013 Materials Research Society
DOI: 10.1557/opl.2013.13

Biosynthesis and Characterization of Bacterial Cellulose Produced by a Wild Strain of Acetobacter spp.

Fatima Yassine[1,2,3], Michael Ibrahim[1], Maria Bassil[1], Ali Chokr[3,4], Anatoli Serghei[2], Antoine El Samrani[3], Mario El Tahchi[1,3] and Gisele Boiteux[2].

[1] LBMI, Department of Physics, Lebanese University - Faculty of Sciences II, PO Box 90656 Jdeidet, Lebanon, email: gbmi@ul.edu.lb, Tel: +961 3 209688, Fax: +961 1 681553.
[2] IMP@LYON1, UMR CNRS 5223 «Ingénierie des Matériaux Polymères», Lyon 1 University, ISTIL Building, 15 Boulevard A. Latarget, 69 622 Villeurbanne Cedex, France,email: gisele.boiteux@univ-lyon1.fr,Tel: +33 4 72 44 85 64, Fax: +33 4 78 89 25 83.
[3] PRASE, Platform of Research and Analysis in Environmental Sciences, Doctoral School of Sciences and Technologies- Lebanese University, PO Box 5, Hadath Campus, Beirut, Lebanon, email: antoineelsamrani@ul.edu.lb Tel: +961 5 470936, Fax.+ 961 5 463278.
[4] Laboratory of Microbiology, Department of Biology, Faculty of Sciences I, Lebanese University, Beirut, Lebanon, e-mail: alichokr@ul.edu.lb, Tel: +961 70 924383.

ABSTRACT

Many advances in nanomaterials synthesis have been recorded during the last 30 years. Bacterial cellulose (BC) produced by bacteria belonging to the genera Acetobacter, Rhizobium, Agrobacterium, and Sarcina is acquiring major importance as one of many eco-friendly materials with great potential in the biomedical field. The shape of BC bulk is sensitive to the container shape and incubation conditions such as agitation, carbon source, rate of oxygenation, electromagnetic radiation, temperature, and pH. The challenge is to control the dimension and the final shape of biosynthesized cellulose, by the optimization of culture conditions. The production of 3D structures based on BC is important for many industrial and biomedical applications such as paper and textile industries, biological implants, burn dressing material, and scaffolds for tissue regeneration. In our work, wild strains of *Acetobacter* spp. were isolated from homemade vinegar then purified and used for cellulose production. Four media of different initial viscosity were used. Cultures were performed under static conditions at 29°C, in darkness. The dimensions and texture of obtained bacterial cellulose nanofibers were studied using scanning electron microscopy (SEM). X-ray diffraction (XRD) showed that the biosynthesized material has a cellulose I crystalline phase characterized by three crystal planes. fourrier transform infrared spectroscopy (FTIR) data confirmed the chemical nature of the fibers. Thermo-gravimetric analysis (TGA) showed that BC preserves a relatively superior non-degradable fraction compared to microcrystalline cellulose.

INTRODUCTION

Eco-friendly materials are acquiring major importance as renewable source with low cost of production. Cellulose is a linear beta 1, 4 linked glucose polymer, and one of the most abundant naturally occurring polymers. Bacterial cellulose (BC) is characterized by thermal stability [1] interfering with high purity and biocompatibility [10,11], flexibility, in situ moldability [6,9,10] and high crystallite width [4,8]. BC is produced by varied bacterial species where *Acetobacter xylinum* is the most efficient producer serving up to now as a model organism for BC biosynthesis. Brown was the first to report bacterial cellulose production in 1886. He identified a gelatinous membrane on the top of *Acetobacter* cultures as chemically equivalent to cell-wall cellulose [2,5]. Later in 1954 Herstin and Schramm optimized BC biosynthesis and developed Herstin Shramm (HS) culture media [10]. Hu et al. in 2010 produced sphere-like bacterial cellulose particles from a specific strain of *Acetobacter xylinum* (JCM 9730) by applying different values of shaking and rotation speed. It is proven that rotational speed has an impact on the internal structure and the size of sphere-like BC [13]. Alkaline treatment is the most common treatment to obtain pure BC membranes; it changes the fibers orientation by the reduction of inter-fibrillar hydrogen bonds. Treatment efficiency depends on the type of solution used, temperature and solution concentration. Nishu *et al.* in 1990 showed that treatments by NaOH, NaClO and H_2O_2 increase Young's modulus and tensile strength [14]. The obtained BC membranes usually contain 99% of water, 10% acting as free bulk water and the majority is tightly bound to BC [8]. In addition, drying process applied after treatment affects the fibers morphology and porosity. Solvent exchange before drying prevents material damaging, for that ethanol or other organic solvent is used. Supercritical carbon dioxide and supercritical ethanol are used to obtain BC aerogels with full preservation of porosity and shape [16]. On the other hand, the addition of water-soluble polymers as hemicellulose or carboxymethylcellulose, chitosan or fluorescent dyes to the liquid culture affects the BC structure [3,12].

In this study, *Acetobacter* spp. strains were isolated from homemade apple vinegar. Modified Herstin Shramm culture media of different viscosity were used to induce BC production. BC biosynthesis was monitored during incubation time by the measurement of pH and the BC production rate. BC films were cleaned then dried and analyzed using scanning electron microscopy (SEM), fourrier transform infrared spectroscopy (FTIR), X-Ray diffraction (XRD) and thermo-gravimetric Analysis (TGA).

EXPERIMENT

Modified Herstin Shramm (HS) culture medium is used in all cultures and isolation procedure. It consists of: 20 g/L glucose (Uni Chem, purity 99.5%), 2.7 g/L disodium hydrogen phosphate (AnalaR, purity 99.97%), 0.115 g/L citric acid (AnalaR, purity 99.8%), 5 g/L yeast extract (Himedia), 5 g/L peptone (Himedia), 2.5 g/L NaCl (Himedia) to prepare broth. 14 g/L of agar (Sigma-Aldrich) are added to prepare solid media. pH is adjusted to 6 using hydrochloric acid. All culture media are autoclaved at 121°C degrees for 15 min, 1% of ethanol (AnalaR) is added before pouring the media in the Petri plates to inhibit cellulose (-) mutants. Agar (Sigma-Aldrich) was used as viscosity modifier to prepare four culture media of different viscosity (M0: culture media without agar, M0.5: culture media with 0.5 g/l of agar, M1: with 1 g/l of agar and M1.5: with 1.5 g/l of agar). *Acetobacter* spp. were isolated by culturing 1 mL of homemade apple vinegar in 24 mL HS broth for 3 days at 29°C. Colonies are purified by repeated streaking of the suspension on HS solid media and incubated at 29°C for 1 day. Bacterial stock is prepared by the addition of 1.5 ml of culture to 1.5 ml of autoclaved glycerol then stocked at −80°C for further cultures. Purified strains are induced to produce cellulose by inoculating 1 mL of a pre-culture in 24 mL HS broth contained in 50 mL sterile tubes incubated at 29°C. Incubation time varied from 1 to 30 days and is maintained under darkness. Cellulose membranes are produced on the top of the broth; they are removed gently and washed excessively with distilled water. Some films are treated for one hour in 1% of NaOH at 70°C to eliminate non-cellulosic compounds then cleaned by distilled water 4 times under agitation for 2 hours each time. Hydrogels are dried at 60°C to obtain finally xerogels.

During the BC biosynthesis, pH of culture media was monitored using a pH-meter (pH − 211 HANNA, country). At each measurement, swelled BC hydrogels are weighted to evaluate the rate of BC production in terms of incubation period. Infrared spectroscopy of freeze-dried BC membranes was recorded on a Fourrier Transform Infrared spectrometer (FTIR − 6300

JASCO); samples were grounded with dried potassium bromide powder and compressed to discs before analysis. Thermal gravimetric curves were obtained for dried samples in TGA Q500 V6.7 Build 203 instrument; samples were heated in open alumina standard pans from room temperature to 600°C under 40 mL/min of helium flow at a heating rate of 20°C /min. Another method was used by repeated heating of the sample from 30 to 250°C and holding at 250°C for 60 to 100 minutes. X-Ray Diffraction was measured using a D8 Bruker diffractometer (anticathode cupper emitting X-ray with wave length $\lambda K\alpha = 0.154060$ nm). Range of 2θ between 10 and 60 degrees were chosen to obtain maximum clarity of crystal phase's diffractogram, with leaps each 0.02 degrees and one second were designed for measurement. Collected diffractograms were analyzed by software EVA and Powder diffraction files provided by interactive center for diffraction data ICDD. SEM were done using SERON AIS2100, freeze-dried BC membrane were coated by a thin carbon layer before observation.

DISCUSSION

Acetobacter spp. isolated from homemade apple vinegar was used to synthesize BC membranes. Herstin Shramm media culture was used, M0 corresponds to the culture media without agar, M0.5, M1 and M1.5 correspond to culture media with 0.5, 1 and 1.5 g/L of agar. This biosynthesis activity was monitored through pH and weight measurements of the culture media and the BC films respectively. Isolated strain was identified as *Acetobacter* spp. based on Bergey's Manual of Determinative Bacteriology [15]. Figure 1 shows BC films on the top of the broth after 14 days of incubation. For M0, a film composed of 2 layers was observed on the top of the broth, one of which was pealing from the other, (Figure 1.a) while the other samples have only one layer. BC synthesizing complexes are associated to pores on the bacteria surface where fibers will be excreted. First glycan chain aggregates are elongated from generation sites or terminal complexes (TC) then assembled in microfibrils which form finally a ribbon, BC fibrils aggregate due to hydrogen bonding and Van der Waals forces and form fibers [3,6,9].

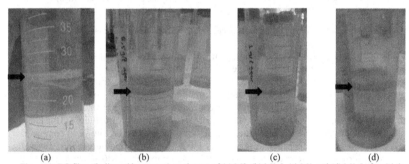

(a)	(b)	(c)	(d)

Figure 1: BC films (indicated by arrows) on the top of (a) M0, (b) M0.5, (c) M1 and (d) M1.5 after 14 days of incubation.

The SEM pictures in Figure 2 show the texture of M0 film sample. The fibers' diameters range from 20 nm to 40 nm.

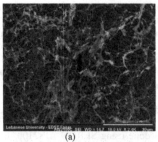
(a)
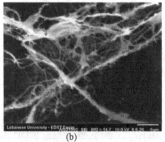
(b)

Figure 2: SEM of BC films collected from the top of M0 medium. The scale bar in (a) is 30 μm and in (b) is 5 μm.

BC production was followed by monitoring the culture media's pH and BC membranes weight. Figure 3 summarizes the variations of pH and rate of cellulose production during the biosynthesis period.

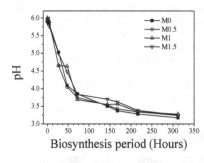
(a)
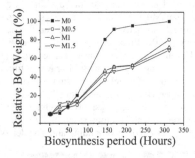
(b)

Figure 3: Variation of (a) pH and (b) BC membranes weight in function of biosynthesis period.

The value of the pH seems to be decreasing of 46% during the biosynthesis (Figure 3.a) explained by the production of acetic acid in the culture media since *Acetobacter* is qualified as acetic acid producer bacterium. Furthermore, Figure 3.b shows that BC membranes weight increases during incubation period and attain its local maximal value at 316 hours. For the first hours of incubation, the media with relatively high concentrations of agar promote the cells to grow on the well-oxygenated surface that is reflected by the high yield of cellulose. But later, this viscosity will disturb the aeration of culture media, which results in the low yield compared to the media prepared without agar. Figure 4.a shows FTIR spectra of microcrystalline fibrous cellulose (Sigma-Aldrich) as reference, non-treated BC membrane from M0 medium and two KOH treated membranes at 72 and 110 hours of incubation period coming from M0 medium as well. The peak at 3450 cm^{-1} corresponds to stretching of hydroxyl groups and is visible in all the samples. While the peak at 2900 cm^{-1} corresponding to C-H stretching of adsorbed water is very weak in BC samples due to the bounded state of water in the BC membrane. The peaks at 1640 cm^{-1} and at 1050 cm^{-1} corresponding to H-O-H bending of adsorbed water and to C-O-C pyranose ring skeletal vibration respectively are visible in all samples, but decreases with treatment [13].

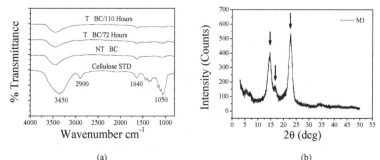

(a)   (b)

Figure 4: (a) FTIR spectra of standard cellulose (cellulose STD), Non-treated BC membrane (NT BC), Treated BC membrane of 72 Hours (T BC/72 Hours) and Treated BC membrane of 110 Hours (T BC/110 Hours). (b) X-ray diffraction patterns of bacterial cellulose film from M1 culture medium after treatment with NaOH.

Figure 4.b presents diffraction pattern taken for BC dry membranes synthesized in the medium M1 with 1 g/L of agar. This spectra represents a profile of cellulose I crystalline phase characterized by three crystal planes −110, 110 and 200 [7]. TGA measurements (Figure 5) showed that BC thermal properties are similar in shape to microcrystalline cellulose standard sample. From Figure 4 it is clear that the degradation of BC and microcrystalline cellulose starts below 320°C. At 400°C fibrous cellulose is totally degraded while BC preserves a relatively superior non degradable fraction, this can be related to the amorphous cellulose fraction present in bacterial cellulose which is more thermal resistant than cellulose I crystalline phase [16].

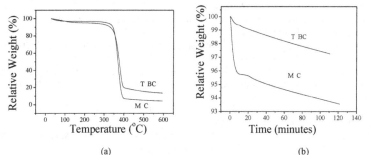

(a)   (b)

Figure 5: (a) TG curves of Treated BC (T BC) and Microcrystalline Cellulose (M C), (b) Hold at 250°C of BC.

In addition, Figure 5.b shows that BC is relatively stable under several cycles of heating under constant temperature (250°C) over 100 minutes period compared to microcrystalline cellulose (MC). Bacterial cellulose seems to retain its initial weight with a loss of about 2.5% after 100 min of treatment while the weight loss of MC is 6.5% of initial weight. This could be explained by the effect of inter H bonds between BC fibers and the higher amount of cellulose Iα confirmation [15] in addition to the high water retention by BC since water is tightly bounded to BC fibers [8].

CONCLUSION

In this study, it is showed that acetic acid bacteria isolated from homemade vinegar is an efficient cellulose producer. FTIR spectra confirm the similarity between bacterial cellulose and microcrystalline cellulose. In addition, it showed that BC structure depends on biosynthesis period and the ulterior treatment. Increasing the viscosity of the medium improved BC yield just in the first days of biosynthesis but seems to inhibit this biosynthesis in the final stages. SEM pictures revealed the random distribution of the nanofibers in the network, while TGA measurements showed important thermal resistance of BC compared to MC during heating treatments. Our study presents a new approach in understanding the behavior of *Acetobacter* spp. during their metabolism in order to control the physical parameters of the produced material.

AKNOWLEDGMENTS

F. YASSINE would like to thank AUF and the Lebanese University for the partial funding of her thesis work. The authors would like to thank Dr. Roland HABCHI for providing SEM pictures. Also the authors appreciate the help of Miss Marwa FOUANI, Miss Sahar RIHAN and Mr. Ali Harb in the microbial, FTIR and DRX analyses. Special thanks to Mr. Olivier GAIN for the TGA measurments and to Mr. Ali HADDANE for the samples preparation.

REFERENCES

[1] A. Bodin, H. Backdahl, H. Fink, L. Gustafsson, B. Risberg, P. Gatenholm, Biotechnol. Bioeng. 97 (2), 425 (2007).
[2] Brown A. J. , J. Chem. Soc 49, 432 (1886).
[3] C. Tokoh, K. Takabe, M. Fujita, H. Saiki, Cellulose (Dordrecht, Neth), 5, 249 (1998).
[4] D. L. Vanderhart, R.H. Atalla, Macromolecules, 17, 1465 (1984).
[5] B. S. Hungund, S. G. Gupta, Asian J.Microbiol. Biotechnol. Environ. Sci, 12, 517 (2010)
[6] Jr RM. Brown, Am Chem Soc, 76 (1992).
[7] K. C. Cheng, J. M. Catchmark, A. Demirci, Journal of Biological Engineering 3, 12 (2009).
[8] K. Gelin, A. Bodin, P. Gatenholm, A. Mihranyan, K. Edwards, M. Stromme Polymer, 48, 7623 (2007).
[9] R. Jonas, L. F. Farah, Polym. Degrad. Stab, 59 (1-3), 101 (1998).
[10] S. Herstin, M. Schramm, Biochem J, 58, 345 (1954).
[11] S. Kobayashi, K. Kashiwa, J. Shimada, T. Kawasaki, S. Shoda, Makromol. Chem., Macromol. Symp., 54/55, 509 (1992).
[12] T. Iwata, L. Indrarti, J-I. Azuma, Cellulose (London), 5 (3), 215 (1998).
[13] Y.Hu, J. Catchmark M., Biomacromolecules, 11, 1727 (2010).
[14] Y. Nishu, M. Uryu, S. Yamanaka, K. Watanabe, N. Kitamura, M. Iguchi, S. Mitsuhashi, Journal of materials Science, 25, 2997 (1990).
[15] D. H. Bergey, G. Holt John, 9 th Edition (Lippincott Williams & Wilkins 1994) p. 71.
[16] F. Liebner, N. Aigner, C. Schimper, A. Potthast, T. Rosenau, ACS Symposium Series: Washington, DC, Chapter 4, 57 (2012).

Mater. Res. Soc. Symp. Proc. Vol. 1498 © 2013 Materials Research Society
DOI: 10.1557/opl.2013.57

Preparation and Characterization of Blends of Polyaniline with Poly(Hydroxybutyrate-Co-Valerate)

David C. da Silva Jr[1], Ana Paula Lemes[1], Lilia M. Guerrini[1], Fernando H. Cristovan[1*]
[1] Universidade Federal de São Paulo, Instituto de Ciência e Tecnologia, São Paulo, Brazil

ABSTRACT

In this study the PANI/PHBV blends were prepared and thermal properties, crystallization behavior, microstructure of the blends were investigated. The PANI/PHBV blends were prepared by dissolution of PANI (emeraldine base doped with dodecylbenzenesulphonic acid, DBSA) and PHBV in chloroform and films were obtained by casting. PANI amount in the blend was varied from 0.1 to 1% wt. PANI/PHBV blends were characterized by FTIR spectroscopy and the thermal behavior were analyzed by differential scanning calorimeter (DSC) and thermogravimetric analysis (TGA). FTIR spectra of the pure PHBV and PANI/PHBV blend had similar peaks. However, blends spectra show an enlargement of bands, due interaction of the chain PANI with PHBV matrix. The crystallization behaviors were investigated using DSC, with at a scanning rate of $10°Cmin^{-1}$. Curve of pure PHBV showed two melting peaks ($159.1°C$ and $172.3°C$). With the increase of PANI amount in the PHBV matrix, both of the melting peaks became wider and shifted to lower temperatures. The decrease trend of first and second melting points with increase of PANI amount, suggests a reduction in the crystallinity of the blends.

INTRODUCTION

Non-biodegradable polymers are extensively used in now a day in the automotive industry, plastic bags and electronic industry. This materials generate pollution problems because remain in the soil for several years. Biodegradable polymers can help overcome reduce petroleum dependency and pollution problems. Among the biodegradable polymers the Polyhydroxyalkanoates (PHA) family is a candidate to solve these problems. A typical example of PHA is poly(hydroxybutyrate-co-hydroxyvalerate) (PHBV), which is produced by bacteria from agricultural raw materials and has good biodegradability, thermoplasticity and biocompatibility [1,2]. PHBV is a derivate of poly-hydroxybutyrate (PHB) the is the most study biodegradable polymer, but often been limited by its narrow processing window and brittleness [3,4], whereas, PHBV is more flexible than PHB due to the presence of the hydroxyvalerate units (HV) in its structure [5]. To improve the properties and range of applications in these materials, the blending of polymers is an alternative way of acquiring a new material with desired properties. There are some studies about a blending of PHBV with semicrystalline polymers such as poly(ε-caprolactone) [6], poly(L-lactic acid) [7], poly(butylene adipate-co-terephthalate) [8]. However, there are not any studies about conducting polymers blends with PHBV. Among conducting polymers the polyaniline (PANI) is the most studied because its unique properties. PANI can be used for development many electronic devices such as sensor [9,10], varistors [11], antistatic protection [12]. For many years it was believed that PANI was not biocompatible, however, it began to change in 2003 with the work of Mattioli-Belmonte et al [13], in which the authors studied the adhesion and proliferation of the cells. The results of this work have shown a great biocompatibility of this material.

The aim of this study was to prepare blends of PANI and PHBV, to investigate the relationship between these two polymers. Evaluate the thermal properties of PANI-DBSA/PHBV blends by differential scanning calorimetry (DSC) and thermogravimetry analysis (TGA). Additionally, structural characterization of the blends was made using UV-Vis and FTIR spectroscopic.

EXPERIMENTAL

PANI in the emeraldine base state (PANI-EB) was chemically synthesized using the standard polymerization process [14]. PANI-EB and PHBV were dissolved apart in a 1:4 v/v m-cresol:chloroform mixture. These mixtures were stirred for 4h at room temperature. Following, dodecylbenzenesulphonic acid (DBSA) was added to the PANI-EB solution. After that it was spirred for 2h. A color change was noticed in this, from dark blue to emerald green, indicating that PANI is already doped by DBSA. Then the PANI-DBSA solution was mixed in the PHBV solution and it was stirred for 4h to blend both polymers. The films were prepared by casting the blend solution over glass plates. Solvent evaporation was performed at 40°C for 24h and the film was easily detached at the end. PANI-DBSA compositions from 0.1 to 1.0% w/w were prepared. UV-Vis spectra of PANI-DBSA/PHBV blend films were acquired using Evolution 220 Thermmo-Scientific UV-Visible Spectrophotometer. FT-IR spectra of samples were obtained using an IRAffinity-1 Shimadzu infrared spectrophotometer. The samples have been through a thermal treatment as well by TGA and DSC. The blend films of PHBV and PANI/PHBV were analyzed in a TGA brand SII Seiko Nanotechnology Exstar model 6000. The weights of the used samples in this test were about 10 mg. The heating was performed from 25°C to 700°C in a 10°Cmin^{-1} rate. The melting temperature (T_f), the melting enthalpy (ΔH_f) and the crystallinity degree ($X_c(\%)$) were measured in a DSC brand DSCII Seiko Nanotechnology e model 6222. Nearly 8mg were heated from 25°C to 240°C in a rate of 10°Cmin^{-1} they kept this temperature for 3 minutes and draughty (to 25°C) having the same rate. The crystallinity percentage was calculated according to the equation 1.

$$X_c(\%) = \frac{\Delta H_f}{\emptyset \Delta_f^\theta} \qquad (1)$$

where:
$\emptyset$ = percentage of PANI in the blend
$\Delta H_f^\theta = 146\ Jg^{-1}$, referring to PHB 100% crystalline [6]

RESULTS AND DISCUSSION

PANI/PHBV were obtained by casting, the films formed showed a good resistance with low amount of PANI (0.1 to 0.5% w/w), but above 0.6% of PANI the films became brittle. The UV-Vis spectra of PANI/PHBV blend films and PHBV film are shown in Figure 1 (a). From the figure we can see that the PHBV practically has no absorption in the region of 1000 at 400 nm. In the 400-300 nm range there is an increase in the absorption is probably from sp^2 carbon. In the PANI/PHBV blend films UV-Vis spectra are observed two bands characteristic of PANI. In both blends films with 0.5 and 0.8 % (w/w) of PANI was observed a weak band in ~340 nm due to the π–π* transition of the benzenoid ring of PANI, in the other film are not observed this absorption because few amount of PANI in the blends. The shoulder in ~450 nm and band in ~770 nm are observed for all the films, this absorptions correspond to polaron π* and π polaron transition [15]. Additionally, was observed a blue shift of 0.8 % PANI/PHBV in relation of 0.1 % PANI/PHBV, this occurs due increase of PANI amount in the blend, what could indicate that PANI is not uniform distributed in PHBV matrix.

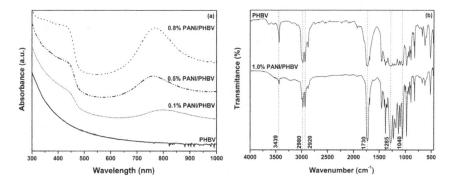

Figure 1. (a) UV-Vis spectra normalized of PHBV pure film and PANI/PHBV blends films with 0.1; 0.5 and 0.8 % (w/w) of PANI. (b) FI-IR spectra of PHBV and PANI/PHBV blends films with 1.0% (w/w) of PANI.

Infrared spectroscopy was used to detect the interaction between PANI-ES and PHBV matrix. Figure 1 (b) the FT-IR transmittance spectra of PHBV film and PANI/PHBV blend film with 1.0 % (w/w) of PANI, in which characteristic peaks of PHBV can be observed. A strong and sharp absorbance band at 1730 cm^{-1} is assigned to the C=O stretching mode of the crystalline parts in PHBV [16]. The weak peaks observed around 1180 cm^{-1}, is attributed to the amorphous state of C–O–C stretching band [17].The band at 3439 cm^{-1} is attributed to its hydroxyl end groups. The bands in 2980 and 2920 cm^{-1} is assigned to the sp^{3} carbons (-C-H stretching vibration). In addition, the peak at 1285 could be assigned to the C-O-C stretching modes of the crystalline parts. In the PANI/PHBV blend spectrum almost cannot be observed peaks of the PANI. However, is observed a small deformation in 3700 at 3000 cm^{-1} perhaps due secondary amine stretch (N-H) of PANI. In the other words, PHBV spectrum is not affected by small amount of PANI. Can be occurs the formation of hydrogen bonds between amine groups of PANI and oxygen of PHBV. Usually, the presence of the absorption bands above 3000 cm^{-1} is possibility of hydrogen bond [18], this fact can explain the blends brightness.

The Table I shows maximum temperatures degradations (T$_{max}$), percentages values of weight losses (W$_1$ and W$_2$) and residues of the blends. The Figure 2(a) shows thermogravimetric curves of the blends. According with this Table, the addition of PANI results in a decrease in degradation temperature. Adding 1.0 % of PANI the degradation process occurs in the temperature of 147°C while in the neat PHBV this process is retarded to temperature of 284°C. This value of maximum temperature degradation of PHBV is in according with the values founded in others studies [19-21]. Also can be verified, according Table I and Figure 2(a), PHBV has a degradation behavior in on step while the blends with 0.1%, 0.5% and 1.0% of PANI showed two degradations steps. In the first step the weight losses were 3.5%, 5.3% and 4.9% between temperatures in the range of 30°C to 207°C. This first step is due the addition of PANI. The degradation behavior of PANI doped with DBSA in three steps has been reported in literature [22,23]. Chen [22] found degradation step in temperature around 100°C and he attributed to evaporation of moisture. The author found weight loss in the second step in temperature around 250°C. This weight loss was attributed to the DBSA evaporation (TGA analysis) with degradation starting at temperature of 200°C. In this work we used DBSA 70% (w/w) in isopropanol and in the cast process was used films the solvent m-cresol. The boiling point of doping agent is 82°C and for the solvent is 203°C [24]. According to Figure 2 (a), the first step is around of 150°C and may be attributed to evaporation moisture and loss of volatiles probably of the doping agent and of the m-cresol. The degradation in the third step of doping PANI with DBSA may occur in the range of 285°C until 550°C [23] or in the range of 150°C until 420°C [22]. In this work the blend of 1.0% of PANI was enough to accelerate the degradation process and the degradation behavior of the PHBV in function of the temperature.

Table I. Maximum temperatures, weight losses, crystallinity degrees, melting temperatures and melting enthalpies of the blends.

PANI (%) in PHBV	T_{max} (°C)	W_1 (%)	W_2(%)	Residue (%)	ΔH_m (mJ/mg)	T_m (°C)	X_c (%)	ΔH_m (mJ/mg)
0	284	----	98.2	1.5	45.2	173	31	45.2
0.1	285	3.5	96.1	0.4	50.4	167	34	50.4
0.5	260	5.3	91.0	3.6	47.9	164	33	47.9
1.0	247	4.9	91.7	3.2	38.7	157	27	38.7

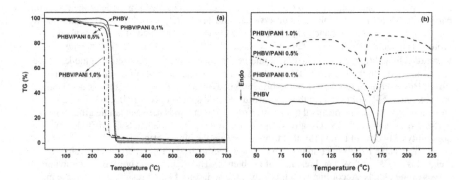

Figure 2. Thermal analyses of PHBV and PANI/PHBV blends with 0.1, 0.5 and 1.0 % of PANI (w/w). (a) TGA curves and (b) DSC curves.

The table I also shows the values of crystallinity degrees (X_c), melting temperatures (T_m) and melting enthalpies (ΔH_m) of the blends. We can be see in Figure 2(b) the thermal behavior of the blends. The PHBV is a copolymer with a semi-crystalline phase (hydroxibutyrate) and an amorphous phase (hydroxyvalerate). This behavior can be observed in the thermogram showed in Figure 2 (b). This Figure shows a narrow and deep peak with a shoulder characterizing the crystalline phase of the polymer. This shoulder is due two peaks together because of the thickness of the crystals and/or recrystallization that occurs during heating in the DSC of the PHBV [17]. According to Table I, the melting temperature of PHBV is 173°C and the melting enthalpy is 45 J/g. The values of melting temperature and enthalpy depend of the hydroxyvalerate degree. The melting temperatures can be in the range from 156°C to 169°C and the melting enthalpy can be in the range of 30 to 64 Jg^{-1} [17,19,25]. In this work the melting temperature and melting enthalpy are in accordance with literature. When PANI is added in PHBV the peak became larger and less deep and the melting temperature decrease. This can be attributed to less perfect crystals. According to Table I, the degree of crystallinity of PHBV is 31%. This value is not altered in the blends with 0.1% and 0.5%, but a little decrease in the blend of PANI (27%). This decrease cannot be considered significant. Besides the melting temperature of PHBV is 173°C and a significant decrease, 157°C, was verified in a 1% PANI blend. The thermal behavior of PANI has been studied [26,27]. Ding et al. [26] verified an endothermic peak between 50°C until 140°C and an extern peak in the range of 185°C until 350°C in the PANI emeraldine base without doping agent. The first peak was attributed to the moisture evaporation and the second peak was verified cross linked reactions. However, in PANI emeraldine base doped with DBSA [27] was observed crystalline phase in temperatures of 121°C and 129°C and glass transitions temperatures (T_g) in the range of 67.3°C (with previous dry) and 76.3°C (without dry). In this work, was not verified thermal transition related to PANI. However, the addition of PANI results in the decrease of the melting temperature, less perfect crystals and no significant decrease in the degree of crystallinity.

CONCLUSIONS

The PANI/PHBV blends were successfully obtained by co-dissolution process. These blends are candidates for biological applications. The blends films formed showed a good resistance with low amount of PANI. UV-Vis and FT-IR spectra showed the interaction between PANI and PHBV. The addition of PANI results in the decrease of the melting temperature, less perfect crystals and no significant decrease in the degree of crystallinity. PHBV has a degradation behavior in one step while the blends with 0.1%, 0.5% and 1.0% showed two degradations steps and the first step is due the addition of PANI. The first step is around of 150°C and may be attributed to evaporation moisture and loss of volatiles, probably of the doping agent and of the m-cresol. The addition of PANI was enough to accelerate the degradation process and the degradation behavior of the PHBV in function of temperature.

ACKNOWLEDGMENTS

Special thanks to FAPESP (2011/17475-6), CAPES and CNPq for the financial support.

REFERENCES

[1] Q.S. Liu, M. Zhu, W.H. Wu, Z.Y. Qin, *Polym. Degr. Stability* **94**, 18 (2009).
[2] Q. Wu, Y. Wang, Artif. Cell. *Blood Sub.* **37**, 1 (2009).
[3] G. E. Luckachan, C.K.S. Pillai, *J. Polym. Environ.* **19**, 637 (2011).
[4] Y. Tokiwa, B. P. Calabia, *J. Polym. Environ.* **15**, 259 (2007).
[5] F. A. Camargo, L. H. Innocentini-Mei1, A. P. Lemos, *J. Compos. Mater.* 1 (2012).
[6] C. D. Gaudio, E. Ercolani, F. Nanni, A. Bianco, *Mater. Sci. Eng. A*, **528**, 1764 (2011).
[7] B. M. P. Ferreira, C. A. C. Zavaglia, E. A. R. Duek, *J. Appl. Polym. Sci.* **86**, 2898 (2002).
[8] A. Javadia, , Y. Srithep, J. Lee, S. Pilla, C. Clemons, S. Gong, L.S. Turng, *Comp.: Part A* **41**, 982 (2010).
[9] D.W. Hatchett, M. Josowicz, *Chem. Rev.* **108**, 746 (2008).
[10] . Janata, M. Josowicz, *Nat. Mater.* **2**, 19 (2003).
[11] F. H. Cristovan, E. C. Pereira, *Synth. Met.* **161**, 2041 (2011).
[12] S. Koul, R. Chandra, S. K. Dhawan, *Polymer* **41**, 9305 (2000).
[13] M. Mattioli-Belmonte, G. Giavaresi, G. Biagini, *Int. J. Artif. Organs.* **26**, 1077 (2003).
[14] F. H. Cristovan, F. R. de Paula, S. G. Lemos, A.J.A. De Oliveira, E. C. Pereira, *Synth. Met.* **159**, 2188 (2011).
[15] S. Stafstrom, J. L. Breadas, A.J. Epstein, H. S. Woo, D.B. Tenner, W.S. Huang, A.G. MacDiarmid, *Phys. Rev. Lett.* **59**, 1464 (1987).
[16] A. Buzarovska, E. A. Grozdanov, *J. Mater. Sci.* **44**, 1844 (2009).
[17] H.Sato, R. Murakami, A. Padermshoke, F. Hirose, K. Senda, I.l. Noda, *Macromolecules* **37**, 7203 (2004).
[18] H. Matsuura, H. Yoshida, M. Hieda, S. Yumanaka, T. Harada, K, Shinya, *J. Am. Chem. Soc.* **125**, 13910 (2003).
[19] M. Avella, M. E. Errico, R. Rimedio, P. Sadocco, *J. Appl. Polym. Sci.* **83**, 1432 (2001).
[20] B. M. P. Ferreira, C. A. C. Zavaglia, E. A. R. Duek, *J. Appl. Polym. Sci.* **86**, 2898 (2002).
[21] L. Shang, Q. Fei, Y.H. Zhang, X. Z. Wang, D. Fan, H. N. Chang, *J. Polym. Env.* **20**, 23 (2012).
[22] C. Chen, *J. Polym. Resear.* **9**, 195 (2002).
[23] M.T. Cortés, V. Sierra. *Polym. Bull.* **56**, 37 (2006).
[24] Site: www.sigmaaldrich.com. Acess: 11/09/12.
[25] C. D. Galdio, E. Ercolani, F. Nanni, A. Bianco, Mat. Sci. Eng. A (2011) 1764.
[26] L. Dining, R.V. Gregory, *Synth. Met.* **104**, 73 (1999).
[27] D. Tsotcheva, T. Tsanov, L. Terlemezyan, S. Vassilev, *J. Therm. Anal. Calorim.* **63**, 133 (2001).

Mater. Res. Soc. Symp. Proc. Vol. 1498 © 2013 Materials Research Society
DOI: 10.1557/opl.2013.86

Synthesis of functionalized-thermo responsive-water soluble co-polymer for conjugation to protein for biomedical applications

Keywords: Protein, synthetic polymer, bioconjugation, hydrogel, tissue engineering, biofactor delivery,

Ali Fathi[1], Hua Wei[1], Wojciech Chrzanowski[2], Anthony S. Weiss[3], Fariba Dehghani[1]

[1] School of Chemical and Biomolecular Engineering, the University of Sydney, NSW 2006, Australia
[2] Faculty of Pharmacy, the University of Sydney, NSW 2006, Australia,
[3] School of Molecular Bioscience, the University of Sydney, NSW 2006, Australia,

ABSTRACT

The aim of this study was to develop a thermo-responsive and bioactive polymer with suitable mechanical properties for musculoskeletal tissue engineering applications. A copolymer was synthesized that comprised of hydrophilic polyethylene glycol, thermo responsive N-isopropylacrylamide (NIPAAm), 2-hydroxyethyl methacrylate-poly(lactide) (HEMA-PLA) to enhance mechanical strength and an active N-acryloxysuccinimide (NAS) group for conjugation to proteins to enhance biological properties. A model protein such as elastin was used to assess the feasibility of conjugating this polymer to protein. The results of ^{1}HNMR analyses confirmed that random polymerization was viable technique for synthesis of this copolymer. The co-polymers synthesized with PEG content of 3 mol% were water soluble. A hydrogel was created by dissolving the copolymer and elastin below room temperature in aqueous media, followed by rapid gelation at 37°C. The results of Fourier transform infrared analyses confirmed the conjugation of protein to copolymer due to significant reduction of ester group absorption (1735 cm^{-1}). This data confirmed molecular interaction between protein and the temperature responsive co-polymer. Our preliminary results demonstrated that it is viable to tune different properties of this hydrogel by changing the composition of co-polymer.

INTRODUCTION

Injectable hydrogels are promising as a minimally invasive approach for tissue engineering [1]. The immense interest to injectable hydrogels is due to their tunable physical and mechanical properties [2–4], and their high water intake capacity [5]. In this approach, the polymer solution is loaded with targeted cells, and injected *in vivo*. It thereafter undergoes solution-to-gel transition (sol-gel) *in situ* due to different chemical or physical stimuli [6–8]. Thermogelation process is an ideal and benign scheme for *in situ* gelation of polymers. In this case a hydrogel is formed when the temperature is shifted to above the lower critical solution temperature (LCST) of thermo-responsive polymer. [9]. Copolymerization of different stimuli-responsive naturally derived polymers such as elastin like polypeptides (ELP) [10–12], gelatin, chitosan [13], hyaluronic acid derived copolymers [14], and synthetic polymers such as poly(N-isopropylacrylamide) (pNIPAAm) [15] has been deemed to be an effective approach to develop polymers with tunable physiochemical, gelation and biochemical properties.

PNIPAAm based copolymers are widely used for synthesis of thermoresponsive injectable hydrogels [15–17]. Application of PNIPAAM based copolymers for tissue engineering however is limited due to lack of cell motif sites in its structure. Natural proteins were mixed with different synthetic polymers to enhance their biological properties [18-19]. Physical mixture of natural and synthetic polymers however might lead to formation of scaffold with non-uniform microstructure. To address this problem, proteins can be chemically conjugated to synthetic polymer to form homogenous hybrid hydrogels. The aim of this study was to synthesize a functionalized, thermoresponsive and water soluble copolymer which can form covalent bonds with proteins.

EXPERIMENTAL
Materials
Chemicals were purchased from Sigma-Aldrich unless otherwise stated. D,L-lactide, stannous 2-ethylhexanoate (Sn(OCt)2), N-isopropylacrylamide (NIPAAm), and N-acryloxysuccinimide (NAS) were used as received. 2-Hydroxyethyl methacrylate (HEMA) was used after distillation under reduced pressure.

Synthesis of HEMA-poly(lactide) (PLA/HEMA) Macromonomer
PLA/HEMA macromonomer was synthesized by ring-opening polymerization of lactide using hydroxyl group of HEMA as initiator and Sn(OCt)$_2$ as catalyst. A mixture of lactide and HEMA were mixed in a three-neck flask. Subsequently, Sn(OCt)$_2$ that was dissolved in anhydrous toluene was added into this mixture. After stirring at 110 °C for 1 h under a nitrogen atmosphere the polymer was purified by precipitation in cold water, centrifugation and drying in several steps.

Synthesis of poly(NIPAAm-co-NAS-co-(HEMA-PLA)-co-OEGMA) (PNPHO)
Poly(NIPAAm-co-NAS-co-(HEMA-PLA)-co-PEG) copolymer, denoted as PNPHO, was synthesized by free radical polymerization. Known amount of monomers were dissolved in tetrahydrofuran and polymerization was conducted for 24 h to acquire desirable molecular weight and yield. The precursors were then precipitated in diethyl ether, filtered, and dried under vacuum.

Elastin-co-PNPHO Hydrogel Formation
Copolymer and elastin were dissolved in aqueous media below room temperature for chemical conjugation. The resulted solution was then kept at 37 °C to form hydrogel. Fourier transform infrared (FTIR) spectroscopy (model Varian 660-IR) was used to confirm the chemical bonding between protein and the copolymer.

RESULTS and DISCUSSION
PLA/HEMA Synthesis
PLA/HEMA macromer was synthesized by ring-opening polymerization of lactide with hydroxyl group of HEMA as initiator and Sn(OCt)$_2$ [20]. The synthesis of PLA/HEMA macromer was confirmed with ^{1}HNMR spectra (Bruker Ultra Shield Advance DPX 400) with evidence of proton peaks from both HEMA and lactyl (LA). Two PLA/HEMA macromers with lactate length of 3 and 6 were synthesized by using 1:1.5 and 1:2.5 initial feed mol ratio of HEMA to LA monomers, respectively.

PNPHO Synthesis
The synthesis of PNPHO copolymer was confirmed with ^{1}HNMR spectra with evidence of proton peaks for each monomer, as shown in Figure 1. It was found that PNPHO copolymer was water soluble, when greater than 3 mol% PEG was used in the synthesis.

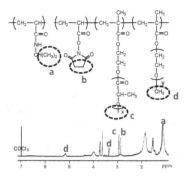

Figure 1. ¹HNMR spectra of PNPHO copolymer and characteristic peaks for each segment (a for NIPAAm, b for NAS, c for PLA/HEMA, and d for PEG based monomer).

Conjugation of PNPHO and Elastin

ATR-FTIR spectra were used to confirm the chemical conjugation between amino group of elastin and activated ester group (succinimide group) of NAS unit in the copolymer. As shown in Figure 2, the copolymer exhibited characteristic peak at 1812 cm^{-1} associated with succinimide group. After the conjugation of elastin, this peak disappeared completely, indicating the participation of elastin in the condensation reaction with succinimide group. It was also observed that the peak for ester group at 1735 cm^{-1} was significantly decreased upon addition of elastin to copolymer. However, the peaks for amide groups at 1630 and 1545 cm^{-1} were increased due to conversion of ester bond to amide linkage during the conjugation of protein with copolymer. These results confirmed the formation of covalent bond between PNPHO and protein which can be used to create bioactive hydrogels with uniform microstructure and suitable mechanical strength.

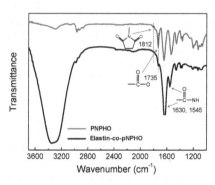

Figure 2. ATR-FTIR spectra of cross-linker (red) and hydrogel (black).

Hydrogel formation

The conjugated product (elastin-*co*- PNPHO) exhibited very low viscosity at room temperature and was injectable through 18G needle. Upon the increase of temperature to 37 °C, the polymer solution converted into hydrogel as shown in Figure 3. The hydrogel exhibited good structural integrity and elastic properties.

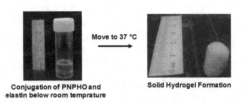

Conjugation of PNPHO and elastin below room temprature

Solid Hydrogel Formation

Figure 3. Formation of hydrogel by conjugation of PNPHO and elastin at 37 °C.

CONCLUSION

A thermoresponsive and water soluble copolymer was synthesized with protein reactive facial groups. This copolymer was chemically conjugated with naturally derived proteins such as elastin. The molecular interaction between elastin and copolymer was confirmed with FTIR analysis. Elastin-copolymer solutions are converted to hydrogels with very high structural integrity by increasing the temperature to 37 °C. This hydrogel might have very high potential for different biomedical applications such as tissue engineering, gene and biofactor delivery.

ACKNOWLEDGMENTS

The authors acknowledge the financial support from the Australian Research Council.

REFERENCES

[1] A. S. Kulshrestha, W. R. Laredo, T. Matalenas, K. L. Cooper, A. Technologies, and R. M. Llc, "Cyclic Dithiocarbonates□: Novel in Situ Polymerizing Biomaterials for Medical," 2010.

[2] N. Annabi, S. M. Mithieux, A. S. Weiss, and F. Dehghani, "Cross-linked open-pore elastic hydrogels based on tropoelastin, elastin and high pressure CO2.," *Biomaterials*, vol. 31, no. 7, pp. 1655–65, Mar. 2010.

[3] N. Annabi, S. M. Mithieux, A. S. Weiss, and F. Dehghani, "The fabrication of elastin-based hydrogels using high pressure CO(2).," *Biomaterials*, vol. 30, no. 1, pp. 1–7, Jan. 2009.

[4] N. Annabi, S. M. Mithieux, E. a Boughton, A. J. Ruys, A. S. Weiss, and F. Dehghani, "Synthesis of highly porous crosslinked elastin hydrogels and their interaction with fibroblasts in vitro.," *Biomaterials*, vol. 30, no. 27, pp. 4550–7, Sep. 2009.

[5] K. Y. Lee and D. J. Mooney, "Hydrogels for Tissue Engineering," *Chemical Reviews*, vol. 101, no. 7, pp. 1869–1880, Jul. 2001.

[6] C. R. Nuttelman, M. a Rice, A. E. Rydholm, C. N. Salinas, D. N. Shah, and K. S. Anseth, "Macromolecular Monomers for the Synthesis of Hydrogel Niches and Their Application in Cell Encapsulation and Tissue Engineering.," *Progress in polymer science*, vol. 33, no. 2, pp. 167–179, Feb. 2008.

[7] S. R. Van Tomme, G. Storm, and W. E. Hennink, "In situ gelling hydrogels for pharmaceutical and biomedical applications.," *International journal of pharmaceutics*, vol. 355, no. 1–2, pp. 1–18, May 2008.

[8] R. Jin, L. S. Moreira Teixeira, P. J. Dijkstra, M. Karperien, C. a van Blitterswijk, Z. Y. Zhong, and J. Feijen, "Injectable chitosan-based hydrogels for cartilage tissue engineering.," *Biomaterials*, vol. 30, no. 13, pp. 2544–51, May 2009.

[9] J. G. Sanchez and T. Tsinman, "Thermoresponsive Platforms for Tissue Engineering and Regenerative Medicine," vol. 57, no. 12, pp. 3249–3258, 2011.

[10] D. L. Nettles, A. Chilkoti, and L. a Setton, "Applications of elastin-like polypeptides in tissue engineering.," *Advanced drug delivery reviews*, vol. 62, no. 15, pp. 1479–85, Dec. 2010.

[11] A. J. Simnick, D. W. Lim, D. Chow, and A. Chilkoti, "Biomedical and Biotechnological Applications of Elastin-Like Polypeptides," *Polymer Reviews*, vol. 47, no. 1, pp. 121–154, Apr. 2007.

[12] D. W. Lim, D. L. Nettles, L. a Setton, and A. Chilkoti, "In situ cross-linking of elastin-like polypeptide block copolymers for tissue repair.," *Biomacromolecules*, vol. 9, no. 1, pp. 222–30, Jan. 2008.

[13] M. Abe, M. Takahashi, S. Tokura, H. Tamura, and A. Nagano, "Cartilage-scaffold composites produced by bioresorbable beta-chitin sponge with cultured rabbit chondrocytes.," *Tissue engineering*, vol. 10, no. 3–4, pp. 585–94, 2004.

[14] N. Iwasaki, Y. Kasahara, S. Yamane, T. Igarashi, A. Minami, and S. Nisimura, "Chitosan-Based Hyaluronic Acid Hybrid Polymer Fibers as a Scaffold Biomaterial for Cartilage Tissue Engineering," *Polymers*, vol. 3, no. 1, pp. 100–113, Dec. 2010.

[15]   J. Guan, Y. Hong, Z. Ma, and W. R. Wagner, "Protein-reactive, thermoresponsive copolymers with high flexibility and biodegradability.," *Biomacromolecules*, vol. 9, no. 4, pp. 1283–92, Apr. 2008.

[16]   D. Neradovic, W. L. J. Hinrichs, J. J. Kettenes-van den Bosch, and W. E. Hennink, "Poly(N-isopropylacrylamide) with hydrolyzable lactic acid ester side groups: a new type of thermosensitive polymer," *Macromolecular Rapid Communications*, vol. 20, no. 11, pp. 577–581, Nov. 1999.

[17]   S. D. Fitzpatrick, M. a Jafar Mazumder, B. Muirhead, and H. Sheardown, "Development of injectable, resorbable drug-releasing copolymer scaffolds for minimally invasive sustained ophthalmic therapeutics.," *Acta biomaterialia*, vol. 8, no. 7, pp. 2517–28, Jul. 2012.

[18]   N. Annabi, A. Fathi, S. M. Mithieux, A. S. Weiss, and F. Dehghani, "Fabrication of porous PCL/elastin composite scaffolds for tissue engineering applications," *The Journal of Supercritical Fluids*, vol. 59, pp. 157–167, Nov. 2011.

[19]   N. Annabi, A. Fathi, S. M. Mithieux, P. Martens, A. S. Weiss, and F. Dehghani, "The effect of elastin on chondrocyte adhesion and proliferation on poly (□-caprolactone)/elastin composites.," *Biomaterials*, vol. 32, no. 6, pp. 1517–25, Feb. 2011.

[20]   W. van Dijk-Wolthuis, "A new class of polymerizable dextrans with hydrolyzable groups: hydroxyethyl methacrylated dextran with and without oligolactate spacer," *Polymer*, vol. 38, no. 25, pp. 6235–42, 1997.

125

Mater. Res. Soc. Symp. Proc. Vol. 1498 © 2013 Materials Research Society
DOI: 10.1557/opl.2013.14

Nanoantibiotic Particles for Shape and Size Recognition of Pathogens

Josef Borovicka[1], Simeon D. Stoyanov[2] and Vesselin N. Paunov*,[1]

[1] Department of Chemistry, University of Hull, Hull, HU6 7RX, UK

[2] Unilever R&D, Olivier van Noortlaan 120, 3133 AT Vlaardingen, the Netherlands.

ABSTRACT

We have developed a novel class of colloidal particles capable of shape and size recognition as well as specific binding to the target cells. These colloid particles were fabricated using a nanoimprinting technology which yields inorganic imprints of the chosen target microorganisms. The products of the templating process are partially fragmented inorganic shells which can selectively bind to their biological counterparts, therefore impairing microbial cell growth, replication and infection. We have named this class of particles, which are capable of selectively recognizing bacterial shape and size, "nanoantibiotics", which can be further functionalized to kill the target cells. The selective binding is driven by the increased area of contact upon recognition of the cell shape and size between the cells and their matching inorganic shell fragments. Here, we demonstrate the cell recognition and binding action of such particles using two different microbial test organisms.

INTRODUCTION

The emergence of bacterial strains resistant event to vancomycin questions our ability to create a universal and lasting defense against novel strains of antibiotic-resistant bacteria[1]. New strategies are needed to confront these growing threats which involve unconventional approaches for containing such microorganisms that use different principle of antibiotic action[2]. One new approach involves nanoparticle formulations with engineered biocidal effect designed to target specific bacteria[3-6]. In another approach Dickert and Hayden[7] used shape recognition of yeast cells by patterned solid surfaces imprinting the surfaces of three different type of yeast, which allowed selective cell binding and distinguishing between them. Similar approached for shape selective binding of microbes and spores on imprinted surfaces and hydrogel beads was employed by Cohen et al.[8] and Harvey et al. [9].

In this paper we have fabricated a new class of the nanoantibiotic colloid particles using a common process of producing silica shells on target cells via the Ströber process of base catalyzed hydrolysis of tetraethoxysilane[10]. Yeast cells were used as model target cells for this templating process. The silica shell deposition was followed by the shell fragmentation using ultrasonic agitation and subsequent bleaching of the yeast cells. After suitable surface treatment, the obtained shell fragments were used to test their binding selectivity in a dispersion of the yeast cells and in a mixture of the same yeast cells and rod-shaped bacterial cells (*B. subtilis*). This methodology is the first step of the fabrication of cell shape-recognizing colloid particles which can be further loaded with highly concentrated biocide that can be delivered directly onto the

---

* Corresponding author. Email: V.N.Paunov@hull.ac.uk, Phone : +44 1482 465660, Fax : +44 1482 466410.

target cell surface. This combination would make these "nanoantibiotic" particles highly efficient selective biocides. In the present work we demonstrate only the selective binding and cell recognition by such nanoparticles. Figure 1 illustrates the fabrication process for these cell shape recognizing nanoantibiotic particles.

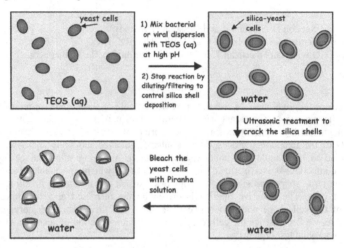

Figure 1. Fabrication scheme of the novel nanoantibiotic particles. The nanoantibiotic particles were fabricated via silica deposition onto yeast cells templates which was followed by the silica shell fragmentation and cell removal via a bleaching process. These inorganic templates of the yeast cells surface were then used to test the shape and size recognition in a dispersion of yeast cells as wells as a mixture of yeast and other bacterial cells.

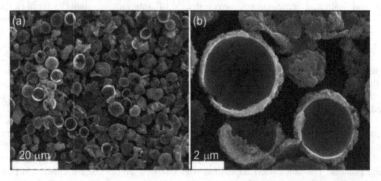

Figure 2. SEM images of the silica shell fragments obtained by templating yeast cells.

RESULTS AND DISCUSSION

The silica shell fragments were produced on yeast cell templates with very good quality. The SEM micrographs of the core/shell particles revealed that some of the produced shell species were already disrupted at the end of the silica deposition procedure. The fragmentation of the silica shells was done using sonication in conjunction with bleaching process, which led to the formation of good quality silica shell fragments which are negative replicas of a part of the yeast cells surface (see Figure 2). These fragmented shells were then utilized in the recognition experiments (Figure 3). The average shell thickness was estimated from the SEM images to be around ~ 220 nm. This thickness, however, depends on the amount of deposited silica and the time of treatment of the target cells.

a) Yeast cells recognition by matching nanoantibiotic particles

The capability of the nanoantibiotic particles to "remember" the size and shape of the target microorganisms was first tested by mixing the shell fragments with their matching yeast cell targets. In these experiments, we also probed the role of the surface chemistry of the nanoantibiotic particles and their matching target cells by coating them with polyelectrolytes. Yeast cells and the silica shell fragments were then incubated together in an aqueous suspension and the cell recognition events were then observed using optical and fluorescence microscopy (Figure 3).

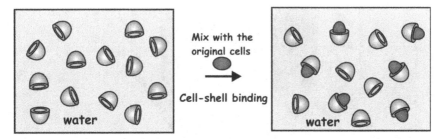

Figure 3. Experimental set up for studying the interaction between yeast cells and their matching silica shell fragments after surface treatment with polyelectrolytes to control their interaction.

The silica shell fragments and the cells were coated with monolayers and bilayers of cationic and anionic polyelectrolytes in order to induce attractive and repulsive electrostatic forces between the matching species. We expected that this would enhance or deter the combination of the negative replicas and their targets. The positive surface charge was induced via the coating of the shell fragments and/or the cells with polyelectrolytes using the layer-by-layer method, where the last coat was done with polyallylamine hydrochloride (PAH), while the negative surface charge was induced using polystyrene sulfonate (PSS) as a last coat. The yeast cells used: (a) were untreated with polyelectrolytes; (b) treated with a monolayer of PAH; or (c) treated with a double layer of PAH and PSS. The silica fragments were (d) left untreated; (e) treated with a monolayer of either PAH or PSS; or (f) treated with a double layer composed of each of those two polyelectrolytes. We tagged the silica shell fragments with Rhodamnine isothiocyanate

(RBITC) while their yeast cell targets were stained with perylene. This was undertaken in order to visualize the recognition incidents as it was realized that the silica shell fragments were often found on the retrograde side of the cell surface with respect to the microscope objective. As expected, we found that the species coated with oppositely charged polyelectrolytes attracted each other and lead to cell recognition by the oppositely charged shell fragments. We also found that the fluorescently tagged silica shell fragments recognized the natively negatively charged yeast cells. Figure 4 contains a graphical summary of the results and a set of sample micrographs.

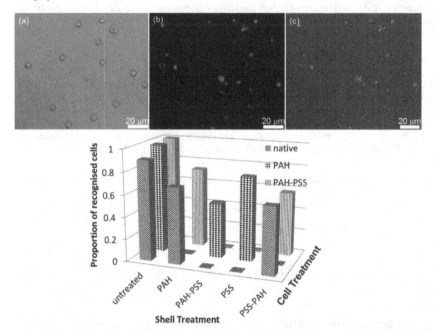

Figure 4. A cell recognition experiment which involved untreated yeast cells and untreated silica shell fragments leading to a match of the silica shell fragments and their cell targets. The bright field image (a) and the fluorescence image (b) were overlaid to produce the image (c). The diagram summarizes the findings from the nanoantibiotic-yeast cell recognition experiments.

In the graphical representation of the results in Figure 4 one sees that the highest rate of recognition interactions between the silica shell fragments and the yeast cell targets was observed for the untreated nanoshells, regardless of the surface coating of the cells. The thickness of the polyelectrolyte coating may also play a certain role in the recognition events resulting in less favorable interactions and a lower recognition rate. It is also expected that other interactions, like van der Waals forces, between the matching silica shell fragments and the target cells, can be strong enough to bind them to their matching cell targets together if their orientation is favorable. We expect that this contribution plays a major effect for untreated cells and non-treated shell fragments. In the case of successful recognition, the shell fragments are attached to their cell counterparts via the concave side which corresponds to the largest achievable area of surface contact with the cell and maximal shell-cell adhesion. Our observations by high resolution optical microscopy also revealed that in the experiments where recognition occurred most of the cells did have attached silica shell counterparts.

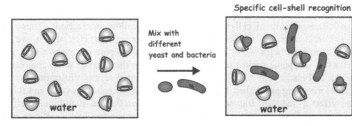

Figure 5. Experimental set up for probing the selective action of nanoantibiotics in a mixture of two types of test microbial organisms of different shape and size.

b) Yeast cells recognition by nanoantibiotic particles in a mixture of bacterial cells

Following the successful demonstration of the specific recognition of individual yeast cells by matching nanoantibiotics, we investigated the nanoantibiotic selectivity in aqueous suspension containing a mixture of two microbial organisms. The nanoantibiotic particles were matching only the shape of the yeast cells whilst the other microbial cells were rod-shaped bacteria (*B. Subtilis*) which are much smaller. Figure 5 illustrates the experimental setup in this case.

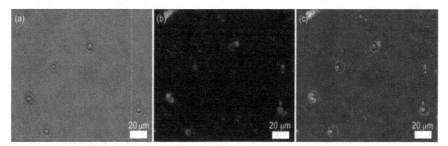

Figure 6. Sample results from the experiments involving the combination of bacterial and yeast cells together with the silica nanoshell fragments designed to match yeast cells. (a) The bright field optical microscopy image and (b) fluorescence microscopy image (b) of the cell mixture incubated with yeast matching nanoantibiotic particles; (c) is the overlay of (a) and (b) which helped us to localize the position of the different cells and the nanoantibiotic particles.

The yeast-templated silica shells fragments were fluorescently tagged with RBITC in the same way as in the experiments involving silica shell fragment-yeast cell interaction (see Figure 6). The recognition was quantified in terms of the percentage of recognized cells and the mismatches. An average $85 \pm 11\%$ recognition of yeast cells with no mismatches was observed. Apart from very small silica fragments with no specific shape, there was no evidence of yeast-templated shell fragments binding to *B. subtilis*.

CONCLUSIONS

We have developed a novel concept for cell shape recognizing colloid particles based on a combination of nanoimprining of cells and shape-specific colloid interactions. These colloid particles are the first step in fabrication of more complex particles which we called "nanoantibiotics". Such particles were fabricated via several preparation steps which involve producing "negative" inorganic replica of the targeted cells in the form of shells fragments that match closely the cell shape and size. We present experimental results illustrating the shape-specific binding of matching silica shell fragments to model target cells like yeast and *B. subtilis* and analyze the effect of the shell fragments surface coating on the cell-shell binding efficiency and the cell recognition. It is anticipated that this novel class of nanoantibiotic particles could be designed to bind shape-specifically and potentially deliver a high dose of biocides directly onto the surface of target bacterial cells. This will allow a single nanoantibiotic particle to deactivate highly antibiotic resistant bacteria where most conventional antibiotics are ineffective. Nanoantibiotic particles can also find applications as non-toxic antibacterial agents, for example to prevent harmful bacteria from growing on home and personal care formulations.

ACKNOWLEDGMENTS

This work was supported by an Industrial CASE studentship funded by BBSRC (grant BB/F01807X/1) and Unilever Research Vlaardingen (The Netherlands).

REFERENCES

1.   C. A. Arias, B. E. Murray, *N. Engl. J. Med.* 2009, 360, 439-443.
2.   D. Abbanat, M. Macielag, K. Bush, *Expert Opin. Investig. Drugs*, 2003, 12, 379-399.
3.   P. Li, J. Li, C. Wu,Q. Wu, J. Li, *Nanotechnology*, 2005, 16, 1912–1917.
4.   H.J. Choi, S.W. Han, S.J. Lee, K. Kim, *J. Colloid Interf. Sci.*, 2003, 264, 458–66.
5.   S. Shrivastava, T. Bera, A. Roy, G. Singh, P. Ramachandrarao, D. Dash, *Nanotechnology*, 2007, 18, 225103.1-9.
6.   F. Martinez-Gutierrez, P. L. Olive, A. Banuelos, E. Orrantia, N. Nino, E. M. Sanchez, F. Ruiz, H. Bach, Y. Av-Gay, *Nanomedicine: Nanotechnology, Biology, and Medicine*, 2010, 6, 681–688.
7.   F. L. Dickert, O. Hayden, *Anal. Chem.* 2002, 74, 1302-1306.
8.   T. Cohen, J. Starosvetsky, U. Cheruti, R. Armon, *Int. J. of Mol. Sci.* 2010, 11, 1236-1252.
9.   S. Harvey, G. Mong, R. Ozanich, J. McLean, S. Goodwin, N. Valentine, J. Fredrickson, *Anal. and Bioanal. Chem.*, 2006, 386, 211-219.
10.   D. Weinzierl, A. Lind, W. Kunz, *Crystal Growth & Design*, 2009, 9, 2318-2323.

Mater. Res. Soc. Symp. Proc. Vol. 1498 © 2013 Materials Research Society
DOI: 10.1557/opl.2013.58

Fructose-Enhanced Efficacy of Magnetic Nanoparticles against Antibiotic Resistant Biofilms

N. Gozde Durmus[1], Erik N. Taylor[1] and Thomas J. Webster[1,2]

[1] School of Engineering, Brown University, Providence, RI, USA 02912
[2] Department of Chemical Engineering, Northeastern University, Boston, MA, USA 02215

ABSTRACT

The emergence of methicillin-resistant *Staphylococcus aureus* (MRSA) is a major cause of hospital-acquired infections (HAI). HAI affect approximately 1.7 million patients each year in the U.S., resulting in up to 100,000 excess deaths, which leads to an estimated cost of more than $35 billion per year. Hence, there is an urgent clinical need to develop new therapies to reduce infections, without resorting to the use of antibiotics for which bacteria are developing a resistance towards. In this study, we designed superparamagnetic iron-oxide nanoparticles (SPION) to treat antibiotic-resistant biofilms and showed that SPION efficacy increases when they are used in combination with fructose.

INTRODUCTION

Biofilms are defined as communities of microorganisms that are adhered to a biotic or abiotic surface.[1, 2] These are complex heterogeneous communities embedded in an extracellular polymeric substance (EPS) matrix, which is composed of polysaccharides, nucleic acids, proteins and lipids.[3, 4] Biofilms are a major source of HAI due to their persistent growth on medical devices and surfaces. The surfaces of invasive medical devices serve as substrates for bacterial attachment, growth and biofilm formation.[5] Biofilm-forming bacteria can withstand host immune responses and become much less susceptible to antibiotics than their individual planktonic counterparts.[2, 6, 7] This causes the antibiotic resistance problem in the clinical settings. Therefore, there is a need to develop novel alternative antibacterial strategies.

Superparamagnetic iron oxide particles (SPION) have been recently used in various biomedical applications for magnetic imaging, bioseperation and biodetection. They are also used for targeted drug delivery applications due to their magnetic properties.[8] Moreover, SPION have shown promising antibacterial properties.[9-12] Moreover, it has been shown that metabolic microenvironment is crucial for antibacterial applications.[13] In this study, our aim was to develop SPION to treat antibiotic-resistant biofilms and to increase SPION efficacy by using metabolic stimulation (i.e., using magnetic nanoparticles in combination with fructose).

EXPERIMENTAL DETAILS

Superparamagnetic iron oxide nanoparticles (SPION) were synthesized using a high temperature synthesis method in triethylene glycol (TREG) and capped with 4 mmol dimercaptosuccinic acid (DMSA). Particle size was analyzed by dynamic light scattering (DLS) measurements. For DLS analysis, SPION were diluted 1:1000 in double deionized water after they were sterile filtered by a 0.2 μm filter. A Zetasizer Nano ZS was used and five readings were taken for each sample.

For bacteria experiments, an antibiotic-resistant strain (MRSA) (ATCC #700699) was used. The bacteria culture was grown in tryptic soy broth (TSB) for 18 hours at 37°C, 200 rpm. Biofilms were established by inoculating 10^8 cells/mL solution in a 96-well plate for 24 hours at 37°C. Then, they were treated with 1 mg/mL of SPION solution. Different concentrations of a fructose solution (1 and 10 mM) were added to investigate their effect on SPION efficacy. After 24 hours of treatment, viable bacteria counts were obtained by disrupting biofilms formed at the bottom of the microtiter plate in a sonicating water bath. First, a 96-well microtiter plate was rinsed three times with 1X phosphate buffer saline (PBS). The plate was placed in a water bath and sonicated for 30 min at 40 kHz to dislodge biofilms into the PBS solution. Then, serial dilutions were performed and 10 μL of each dilution was spot-plated on agar plates. Colony forming units (CFU) were counted after overnight incubation at 37°C.

RESULTS AND DISCUSSION

Material characterization:

SPION morphology characterized by transmission electron microscopy (TEM) is shown in Figure 1. Magnetic nanoparticles were synthesized using a variation of the high temperature reflux method in triethylene glycol (TREG).[14] Dimercaptosuccinic acid (DMSA) chelation is known to increase the solubility and dispersion of particles in water, cell medium and physiological fluids.[15]

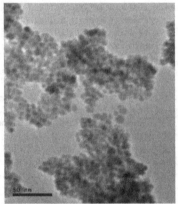

Figure 1. TEM characterization of SPION. TEM analysis demonstrated that SPION were successfully synthesized with an average size of 10.07 ± 1.54 nm. (Scale bar = 50 nm)

TEM analysis demonstrated that SPION were successfully synthesized with an average size of 10.07 ± 1.54 nm. SPION were also characterized by dynamic light scattering (DLS) measurements. According to DLS analysis, magnetic nanoparticles had a mean hydrodynamic diameter of 18.17 ± 1.34 nm (Figure 2).[16] Larger particle size analysis from DLS measurements were expected since dimercaptosuccinic acid (DMSA) was used as a capping agent during magnetic nanoparticle synthesis. Nanoparticle aggregation or coating agents can interfere with the hydrodynamic diameter measurements.[16-18] Therefore, DLS particle size measurements can be different from the nanoparticle size measured by TEM.

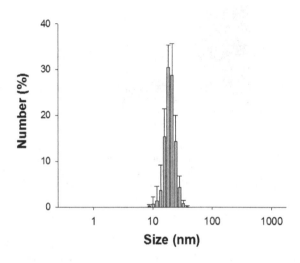

Figure 2. Dynamic light scattering analysis of SPION used in the study. According to DLS analysis, SPION had a mean hydrodynamic diameter of 18.17 ± 1.34 nm.

Assessment of Antibacterial Properties of Magnetic Iron Oxide Nanoparticles:

Quantification of colony forming unit (CFU) analysis after 24 hours of SPION treatment is shown in Figure 3. It is demonstrated that 1mg/mL SPION treatment significantly decreased the number of viable bacteria cells within the biofilm, compared to untreated control biofilms (p < 0.001). Most importantly, antibacterial properties were significantly improved when 10 mM fructose was added to the SPION treatment (p < 0.05).

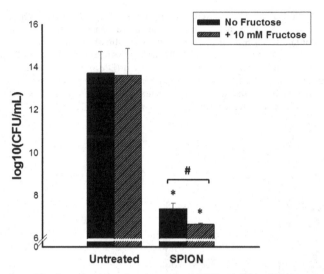

Figure 3. Quantification of colony forming units (CFU/mL) after SPION treatment. SPION significantly reduced the viable cells within the biofilm. Anti-bacterial properties are enhanced when SPION are used in combination with 10 mM fructose. Data represents mean ± SEM. N = 3. *p < 0.001 compared to untreated biofilms; # p < 0.05 compared to the SPION treatment without fructose.

CONCLUSIONS

There is an urgent clinical need for novel biofilm eradication strategies. Here, we present that combining magnetic iron oxide nanoparticles with metabolic stimulation (i.e., providing fructose in the nanoparticle treatment) can further decrease bacterial functions and eradicate antibiotic-resistant biofilms. In addition, our results show the importance of biofilm metabolic microenvironment for nanomedicine applications. SPION-based tretments can be targeted to the infection site and eradicate the biofilm, in the presence of a magnetic field.[17] Thus, these novel treatments have the potential to be used as an alternative strategy to antibiotics.

ACKNOWLEDGMENTS

The authors would like to thank Hermann Foundation and Center for Integration of Medicine and Innovation (CIMIT) Prize for Technology in Primary Healthcare for funding.

REFERENCES

[1]   G. O'Toole, H. B. Kaplan, and R. Kolter, "Biofilm formation as microbial development," *Annual Review of Microbiology*, vol. 54, pp. 49-79, 2000.

[2]   L. Hall-Stoodley, J. W. Costerton, and P. Stoodley, "Bacterial biofilms: From the natural environment to infectious diseases," *Nature Reviews Microbiology*, vol. 2, pp. 95-108, Feb 2004.

[3]   Xavier JB, Picioreanu C, Rani SA, van Loosdrecht MC, and Stewart PS., "Biofilm-control strategies based on enzymic disruption of the extracellular polymeric substance matrix-a modelling study.," *Microbiology*, vol. 151, pp. 3817-32., 2005.

[4]   Davey ME and O'toole GA., "Microbial biofilms: from ecology to molecular genetics," *Microbiol Mol Biol Rev.*, vol. 64, pp. 847-67, 2000.

[5]   Antoci V Jr, Adams CS, Parvizi J, Davidson HM, Composto RJ, Freeman TA, Wickstrom E, Ducheyne P, Jungkind D, Shapiro IM, and Hickok NJ, "The inhibition of Staphylococcus epidermidis biofilm formation by vancomycin-modified titanium alloy and implications for the treatment of periprosthetic infection.," *Biomaterials*, vol. 35, pp. 4684-90, 2008.

[6]   D. G. Davies, A. M. Chakrabarty, and G. G. Geesey, "EXOPOLYSACCHARIDE PRODUCTION IN BIOFILMS - SUBSTRATUM ACTIVATION OF ALGINATE GENE-EXPRESSION BY PSEUDOMONAS-AERUGINOSA," *Applied and Environmental Microbiology*, vol. 59, pp. 1181-1186, Apr 1993.

[7]   Costerton JW, Stewart PS, and Greenberg EP., "Bacterial biofilms: a common cause of persistent infections.," *Science*, vol. 284, pp. 1318-22., 1999.

[8]   R. Hao, R. J. Xing, Z. C. Xu, Y. L. Hou, S. Gao, and S. H. Sun, "Synthesis, Functionalization, and Biomedical Applications of Multifunctional Magnetic Nanoparticles," *Advanced Materials*, vol. 22, pp. 2729-2742, Jul 2010.

[9]   N. Tran, A. Mir, D. Mallik, A. Sinha, S. Nayar, and T. J. Webster, "Bactericidal effect of iron oxide nanoparticles on Staphylococcus aureus," *International Journal of Nanomedicine*, vol. 5, pp. 277-283, 2010.

[10]   D. Lee, R. E. Cohen, and M. F. Rubner, "Antibacterial properties of Ag nanoparticle loaded multilayers and formation of magnetically directed antibacterial microparticles," *Langmuir*, vol. 21, pp. 9651-9659, Oct 2005.

[11]   E. N. Taylor and T. J. Webster, "The use of superparamagnetic nanoparticles for prosthetic biofilm prevention," *International Journal of Nanomedicine*, vol. 4, pp. 145-152, 2009.

[12]   H. Park, H. J. Park, J. A. Kim, S. H. Lee, J. H. Kim, J. Yoon, and T. H. Park, "Inactivation of Pseudomonas aeruginosa PA01 biofilms by hyperthermia using superparamagnetic nanoparticles," *Journal of Microbiological Methods*, vol. 84, pp. 41-45, Jan 2011.

[13]   Allison KR, Brynildsen MP, and Collins JJ., "Metabolite-enabled eradication of bacterial persisters by aminoglycosides," *Nature*, vol. 73, pp. 216-20, 2011.

[14]   J. Wan, W. Cai, X. Meng, and E. Liu, "Monodisperse water-soluble magnetite nanoparticles prepared by polyol process for high-performance magnetic resonance imaging," *Chemical Communications*, pp. 5004-5006, 2007.

[15]   W. Yantasee, C. L. Warner, T. Sangvanich, R. S. Addleman, T. G. Carter, R. J. Wiacek, G. E. Fryxell, C. Timchalk, and M. G. Warner, "Removal of heavy metals from aqueous

systems with thiol functionalized superparamagnetic nanoparticles," *Environmental Science & Technology,* vol. 41, pp. 5114-5119, Jul 2007.

[16]   R. C. Murdock, L. Braydich-Stolle, A. M. Schrand, J. J. Schlager, and S. M. Hussain, "Characterization of nanomaterial dispersion in solution prior to In vitro exposure using dynamic light scattering technique," *Toxicological Sciences,* vol. 101, pp. 239-253, Feb 2008.

[17]   Durmus NG and Webster TJ, "Eradicating Antibiotic-Resistant Biofilms with Silver-Conjugated Superparamagnetic Iron Oxide Nanoparticles," *Advanced Healthcare Materials,* 2012.

[18]   M. A. Dobrovolskaia, A. K. Patri, J. W. Zheng, J. D. Clogston, N. Ayub, P. Aggarwal, B. W. Neun, J. B. Hall, and S. E. McNeil, "Interaction of colloidal gold nanoparticles with human blood: effects on particle size and analysis of plasma protein binding profiles," *Nanomedicine-Nanotechnology Biology and Medicine,* vol. 5, pp. 106-117, Jun 2009.

Mater. Res. Soc. Symp. Proc. Vol. 1498 © 2013 Materials Research Society
DOI: 10.1557/opl.2013.160

AN ENZYMATIC METHOD TO OBTAIN A NEW SCAFFOLD FOR ENGINEERING CARTILAGE

Giraldo Gomez David M.[1], Villegas Alvarez Fernando[2], Garciadiego Cazares David[3], Sotres Vega Avelina[4], Piña Barba Maria C.[1]

[1]Instituto de Investigaciones en Materiales, UNAM, Ciudad Universitaria, Circuito Exterior s/n, C.P. 04510, México D.F., México.
[2]Departamento de Cirugía-Facultad de Medicina, UNAM, Ciudad Universitaria, Circuito Exterior s/n, C.P. 04510, México D.F., México.
[3]Unidad de Ingeniería de Tejidos Terapia Celular y Medicina Regenerativa, Instituto Nacional de Rehabilitación, Calzada México Xochimilco N° 289, C.P. 14389, México D.F., México.
[4]Departamento de Investigación en Cirugía Experimental, Instituto Nacional de Enfermedades Respiratorias "Ismael Cossio Villegas", Calzada de Tlalpan N° 4502, C.P. 14080, México D.F., México

ABSTRACT

The purpose of this study was to achieve a descellularized scaffold from cartilage tissue, which can be used as xenograft for cartilage tissue regeneration.

This work presents the results obtained using one method to wash porcine trachea in order to remove cellular material from the extracellular matrix and to avoid the immune reaction using enzymatic detergent and partial enzymatic degradation with Deoxyribonuclease I (DNase-I), Ethylenediaminetetraacetic Acid (EDTA) and Trypsin. This treatment was qualitatively evaluated by Scanning Electron Microscopy (SEM), and H&E Stain (Histology), and quantitatively evaluated by DNA quantification. The thermal characterization of the descellularized scaffold was carried out using Termogravimetric Analysis (TGA) and Differential Scanning Calorimetry (DSC). The type of collagen obtained from the scaffold was determined through SDS-PAGE electrophoresis. When using Enzymatic Treatment (ET) to wash trachea tissue, it is possible to obtain an acellular xenograft; this procedure has the potential to avoid rejection reactions of the xenograft.

INTRODUCTION

The self-regeneration of cartilage is a very complicated process, this due to its low regenerative capacity and its low blood provision. However tissue engineering using biodegradable scaffolds and cell replacement therapy has become one of the promising approaches in the regeneration of cartilage [1-5]. The scaffolds provide a 3D structure that enables the cell growth while the differentiation function occurs. Numerous materials have been used to build these scaffolds, among which are those of natural origin as polylactic acid (PLA) polyglycolic acid (PGA), collagen type I, and descellularized cartilage matrix [2-10]. The descellularized matrices obtained from cartilage (ECM), which contain the naturals elements of cartilage with good biocompatibility for the growth of chondrocytes [11].

Theoretically, these scaffolds can be ideal for use in cartilage engineering, however, the used process to decellularization of the tissue is crucial, since all cellular material must be removed to avoid any type of immune response post-implantation, either way, the structural components from the ECM should be preserved to provide a scaffold with good biomechanical features and promote efficient invasion of tissue cells [12]. The differences in the structures of

ECM makes difficult the process of removing cells with minimal tissue destruction. Given these considerations, descellularized matrices were obtained from cartilage tissue, such as the trachea [13-15], resulting in a favorable immune response when implanted as xenografts or allografts [16]. Using these results it has been obtained scaffolds for use in the replacement of human trachea, with good results in clinical trials [1,17].

This paper presents a tracheal decellularization process using enzymatic detergents.

MATERIALS AND METHODS

Obtaining matrix

We obtained a 15cm section of trachea from an animal swine young and healthy, according to NOM-062-ZOO-1999 [18].

Descellularization process

To carry out the process of decellularization, the trachea was cut in three portions of 5cm in length, the fibro connective tissue of them was removed using a scalpel.

The first segment was subjected to 5 cycles of wash with an enzymatic treatment (TE), the cycle starts with a rinse using a preliminary yodopovidone 1% in PBS and was continued with washings in an antibiotic-antimycotic solution at 1% in Milli-Q water, after was placed in 10mM EDTA (Sigma-Aldrich) for 3h. After rinsing the tissue with deionized water was placed for another 3 h in Trypsin (Sigma-Aldrich) to 1% (W/V) in PBS solution, then was placed in sodium deoxycholate (Sigma-Aldrich) for 4 hours, pre-rinsing with deionized water and was continued with another rinsing with deionized water and then was placed in DNase-I (invitrogen) in a hypotonic solution for 4 hours. Finally was rinsed in deionized water and was stored after complete 5 cycles in a solution of PBS with antibiotic-antimycotic [19,20].

The second segment of the trachea was treated according to the detergent-enzymatic method (DEM) [17], 15 washed cycles were carried out.

The third segment tracheal was treated according to the method DEM, adding an incubation cycle in ethylene diamine tetra acetic acid (EDTA), for latest treatment were carried out 15 cycles of decellularization [19,20].

CHARACTERIZATION

Histology

The tissues were subjected to a gradual dehydration ethyl alcohol starting with 70% and they finished with absolute alcohol, the tissue was fixed with xylene and embedded in paraffin. Slices of 5 μm were cut with the aid of a rotary microtome (Leica) the samples were stained with hematoxylin and eosin (H & E) (Merck) [19].

Scanning electron microscopy

This analysis was done to assess qualitatively the decellularization process and to observe the structure of the matrix. The samples were fixed using 3% (v/v) buffered glutaraldehyde in 0.1M sodium cacodylate (pH 7.2), after were proceeded to dehydration using gradient of ethanol method and finally were dried using a CO_2 chamber in critical point [19].

DNA quantification

To measured the effectiveness of this method, the DNA from cells of the samples was isolated employing TRIzol ® (Invitrogen) [21-22] and then it was measured the DNA concentration using an absorbance spectrophotometer (GeneQuant) [19].

SDS-PAGE electrophoresis

The existent proteins in the matrix after process were characterized with this technique; also it is possible to identify if there was any protein degradation. The proteins were quantified by the Lowry method; a standard Bovine Serum Albumin (BSA, fraction V, Calbiochem) for the calibration curve was used [19].

Differential Scanning Calorimetry

This analysis was made with the aim to characterize the thermal behavior of macromolecules as well as to confirm if there were molecular degradations in the matrix after the decellularization process. For this analysis the samples were lyophilized. A differential scanning calorimeter (DSC) (TA Instruments Q100) at heating rate of 10°C/min in N atmosphere in a T range from 25°C to 300°C was used [19].

Thermogravimetric Analysis

This study was made using a Thermogravimetric Analyzer (TGA) (TA Instruments Q500) with a heating rate of 10°C/min. Under Nitrogen atmosphere using a temperature range from 25°C to 500°C [19].

RESULTS AND DISCUSSION

Microscopic evaluation

After performing the process of decellularization, the histological analysis of different samples revealed that in the case of the native sample (untreated) there were chondrocytes embedded in the lacunae of the cartilage matrix, as expected. The samples treated with the DEM process still showed the presence of cells, since this treatment is only effective for complete decellularization after 25 cycles. The cells were still presented in the treatment with DEM + EDTA, because the addition of chelating agent didn't had a significant effect in their reduction. Finally, for the sample subjected to TE treatment, the complete decellularization of the matrix was evident. (Figure 1).

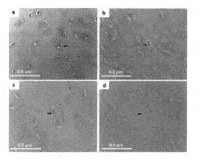

Fig. 1. H&E stain trachea cartilage showing the presence (a) of cellular elements in native trachea. (b) after 15 **DEM** cycles and (c) 15 **DEM** cycles treatment with addition of EDTA, cartilage matrices showing the presence of cellular elements but without nucleus of cells, and (d) H&E stain after 5 cycles of **ET** showing the absence of any cellular element.

With the histological results for each process, and considering different sections from each sample, the quantification of cells was performed. From these, it was possible to obtain a quantitative result of each process, as shown in Figure 2.

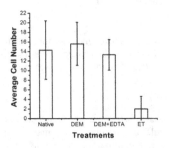

Fig. 2. Average cell number quantification for native sample and the three different descellularization treatments, from the histology fields. The average cell number shows there is not significantly different when compare the cell number of the native with DEM and DEM+EDTA treatment otherwise, the ET shows a evident decrease in the average cell number.

In the micrographs, a proper morphology and little changes of the architecture of the matrices in all cases were observed, as shown in Figure 3.

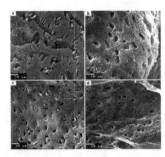

Fig. 3. (a) SEM micrograph of native trachea sample showing the presence of cellular elements embodiment in the cartilage matrix. (b) SEM micrograph after 15 **DEM** cycles and (c) 15 **DEM** cycles with addition of EDTA, tracheal matrices showing the presence of few cellular elements embodiment in the cartilage matrix and (d) SEM micrograph after 5 cycles of enzymatic treatment **ET** showing the absence of any cellular element and a good aspect of the cartilage matrix without appearance alteration.

DNA quantification

This quantification is an indirect measure to confirm the effectiveness of the processes, since the DNA is into active nucleus of cell; this result is shown in Figure 4.

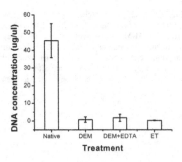

Fig. 4. DNA quantification for native sample and the three different descellularization treatments, this result is due to the presence of the DNase-I enzyme in the three treatments, this explain why in the DEM and DEM+EDTA treatments, the cells appear without nucleus in the histology characterization.

The results of microscopy, as well as the quantification of DNA, confirmed a decrease in the number of cells in the matrix, which involves the destruction of the cellular component and a decrease of the immune response for the matrix.

Electrophoresis

The electrophoresis patterns obtained were typical of collagen II, bands were observed α1 (II) to a weight 110 kD and the β-chains were found around 250 kD, as shown in Figure 5.

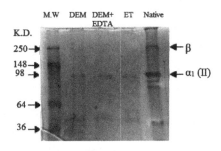

Fig. 5. Gel SDS-PAGE electrophoresis, (M.W) Molecular Weight marker *SeeBlue* (BIO-RAD) and three different treatments with native cartilage patterns, , the gel shows a typical Collagen II pattern, no alterations or degradation were observed

Thermal Analysis

Three measurements were made for each decellularization process, using DSC and TGA. They determined the denaturation temperature (Td), degradation temperature and whether alteration of these parameters is observed. Figures 6 and 7 show the average of the measurements obtained.

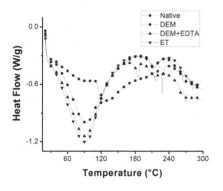

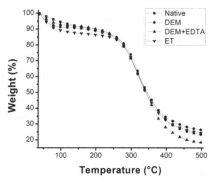

Fig. 6. DSC profile of native cartilage and treatments

Fig. 7. TGA profile of native cartilage and treatments

The results of thermal analysis, as well as electrophoresis, showed that the treatment had no adverse effect on the structural characteristics of the matrix.

CONCLUSIONS

With the experimental methodology used in this work, it was possible to obtain a fully decellularized matrix (ET) as shown the microscopic evaluation and DNA quantification in shorter time than the previously reported for these matrices. Even though the amount of DNA was reduced in the three treatments only in the ET was not observe any other cellular element. The characterization does not show any alteration in the architecture nor the molecular structure of the matrix. This matrix presents the potential to be used as a scaffold in tissue engineering.

ACKNOWLEDGMENTS

To DGAPA –UNAM for the financial support through IT104011 project. To Dr. Ignacio Figueroa, Dr. Omar Novelo, B.Sc. Esteban Fregoso, B.Sc. Armando Zepeda, Dr. Ericka Peña and Dr. Lizeth Fuentes, for their technical support.

REFERENCES

1. Macchiarini P, Jungebluth P, Go T, Asnaghi MA, Rees LE, Cogan TA, et al. Clinical transplantation of a tissue-engineered airway. Lancet 2008; **372** (9655): 2023-30.
2. Cao Y, Vacanti JP, Paige KT, Upton J, Vacanti CA. Transplantation of chondrocytes utilizing a polymer-cell construct to produce tissue-engineered cartilage in the shape of a human ear. Plast Reconstr Surg 1997; **100** (2): 297-302.
3. Ciorba A, Martini A. Tissue engineering and cartilage regeneration for auricular reconstruction. Int J Pediatr Otorhinolaryngol 2006;**70**(9): 1507-15.
4. Vacanti CA, Kim W, Schloo B, Upton J, Vacanti JP. Joint resurfacing with cartilage grown in situ from cell-polymer structures. Am J Sports Med 1994; **22**(4): 485-8.
5. Gong YY, Xue JX, Zhang WJ, Zhou GD, Liu W, Cao Y. A sandwich model for engineering cartilage with acellular cartilage sheets and chondrocytes. Biomaterials 2011; **32**: 2265-2273.
6. Hui TY, Cheung KM, Cheung WL, Chan D, Chan BP. In vitro chondrogenic differentiation of human mesenchymal stem cells in collagen microspheres: influence of cell seeding density and collagen concentration. Biomaterials 2008; **29**(22): 3201-12.
7. Schmal H, Mehlhorn AT, Kurze C, Zwingmann J, Niemeyer P, Finkenzeller G, et al. In vitro study on the influence of fibrin in cartilage constructs based on PGA fleece materials. Orthopade 2008; **37**(5): 424-34.
8. Piña M.C., Romero M., Tello S., Labastida A., Dávalos K., Rosales D., Fregoso E., Colágena tipo I: Obtención y caracterización, XX Congreso Nacional de la Sociedad Polimérica de México, Guanajuato, México, 30 Oct. Al 2 de Nov. del 2007 pp.459-463.
9. Gómez Lizárraga K., Piña Barba C., Rodríguez Fuentes N., Romero M., Obtención y caracterización de colágena tipo I a partir de tendón de bovino, Superficies y Vacío, Soc. Mex. de Ciencia y Tecnología de Superficies y Materiales 2011: **24**(4), 137-140.
10. María Luisa Del Prado Audelo, María del Carmen García de León Méndez, Cristina Piña Barba, Extracción de colágena tipo I de distintos tejidos biológicos y su caracterización, International Conference on Polymers and Advanced Materials POLYMAT-2011, Huatulco, México. 16-21 Octubre 2011.
11. Conconi MT, De Coppi P, Di Liddo R, Vigolo S, Zanon GF, Parnigotto PP, et al. Tracheal matrices, obtained by a detergent-enzymatic method, support in vitro the adhesion of chondrocytes and tracheal epithelial cells. Transpl Int 2005; **18**(6): 727-34.
12. Gilbert TW, Sellaro TL, Badylak SF. Decellularization of tissues and organs. Biomaterials 2006;**27**: 3675-83.
13. Macchiarini P, Walles T, Biancosino C, Mertsching H. First human transplantation of a bioengineered airway tissue. J Thorac Cardiovasc Surg 2004; **128**: 638-41.
14. Walles T, Giere B, Hofmann M, Schanz J, Hofmann F, Mertsching H, et al. Experimental generation of a tissue-engineered functional and vascularized trachea. J Thorac Cardiovasc Surg 2004;**128**: 900-6.
15. Conconi MT, De Coppi P, Bellini S, Zara G, Sabatti M, Marzaro M, et al. Homologous muscle acellular matrix seeded with autologous myoblasts as a tissue-engineering approach to abdominal wall-defect repair. Biomaterials 2005; **26**: 2567-74.
16. Jungebluth P, Go T, Asnaghi A, Bellini S, Martorell J, Calore C, et al. Structural and morphological evaluation of a novel enzymatic detergent tissue engineered tracheal tubular matrix. J Thorac Cardiovasc Surg 2009; **138**: 586-93.
17. Baiguera S, Jungebluth P, Burns A, Mavilia C, Haag J, De Coppi P, Macchiarini P. Tissue engineered human tracheas for in vivoimplantation. Biomaterials 2010; **31**: 8931-8938.
18. Especificaciones Técnicas para la Producción Cuidado y Uso de Animales de Laboratorio de la Norma Oficial Mexicana NOM-062-ZOO-1999. Diario Oficial de la Federación 1999:Diciembre 6. Estados Unidos Mexicanos.
19. Giraldo Gomez David M. Tesis de Maestría. "Obtención de un andamio acelular para substitución de tráquea". Universidad Nacional Autónoma de México, Diciembre, 2011.
20. Giraldo David Mauricio, Piña Maria Cristina, Villegas Fernando, Sotres Avelina. New Scaffold from cartilage tissue allows reproduction of trachea, International Conference on Polymers and Advanced Materials POLYMAT-2011, Huatulco, México. 16-21 Octubre 2011. Oral S1-38.
21. Chomczynski, P., Sacchi, N. Anal. Biochem. 1987;**162**, 156.
22. Chomczynski, P. Biotechniques.1993;**15**, 532.

Mater. Res. Soc. Symp. Proc. Vol. 1498 © 2012 Materials Research Society
DOI: 10.1557/opl.2012.1653

Structure-property-function relationships in triple-helical collagen hydrogels

Giuseppe Tronci,[1,2] Amanda Doyle,[1,2] Stephen J. Russell,[2] and David J. Wood[1]

[1]Biomaterials and Tissue Engineering Research Group, Leeds Dental Institute, University of Leeds, Leeds LS2 9LU, United Kingdom
[2]Nonwoven Research Group, Centre for Technical Textiles, University of Leeds, Leeds LS2 9JT, United Kingdom

ABSTRACT

In order to establish defined biomimetic systems, type I collagen was functionalised with 1,3-Phenylenediacetic acid (Ph) as aromatic, bifunctional segment. Following investigation on molecular organization and macroscopic properties, material functionalities, i.e. degradability and bioactivity, were addressed, aiming at elucidating the potential of this collagen system as mineralization template. Functionalised collagen hydrogels demonstrated a preserved triple helix conformation. Decreased swelling ratio and increased thermo-mechanical properties were observed in comparison to state-of-the-art carbodiimide (EDC)-crosslinked collagen controls. Ph-crosslinked samples displayed no optical damage and only a slight mass decrease (~ 4 wt.-%) following 1-week incubation in simulated body fluid (SBF), while nearly 50 wt.-% degradation was observed in EDC-crosslinked collagen. SEM/EDS revealed amorphous mineral deposition, whereby increased calcium phosphate ratio was suggested in hydrogels with increased Ph content. This investigation provides valuable insights for the synthesis of triple helical collagen materials with enhanced macroscopic properties and controlled degradation. In light of these features, this system will be applied for the design of tissue-like scaffolds for mineralized tissue formation.

INTRODUCTION

Collagen is the main protein of the human body, ruling structure, function and shape of biological tissues. Also in light of its unique molecular organization, collagen has been widely applied for the design of vascular grafts [1], fibrous materials for stem cell differentiation [2], biomimetic scaffolds for regenerative medicine [3], and tissue-like matrices for hard tissue repair [4]. However, collagen properties are challenging to control in physiological conditions, mainly because its hierarchical organization and chemical composition *in vivo* can only be partially reproduced *in vitro*. Functionalisation and crosslinking of collagen molecules, e.g. via carbodiimide [5,6], glutaraldehyde [7,8] or hexamethylene diisocyanate [9], have proved to enhance macroscopic properties in aqueous environment, although much is still left to do to establish biomimetic systems with defined structure-property-function relationships. Here, the design of type I collagen hydrogels was investigated via covalent lysine functionalisation with 1,3-Phenylenediacetic acid (Ph). It was hypothesized that incorporation of a stiff, aromatic segment among collagen molecules could offer a novel synthetic route to the formation of mechanically-relevant materials. Ph was selected as bifunctional segment, in order to promote crosslinking of distant collagen molecules, unlikely accomplished with current synthetic methods [1,5,6], so that triple helix conformation of collagen could be retained. The presence of an aromatic ring in the Ph was considered crucial to achieve controlled swelling and enhanced mechanical properties in resulting hydrogels, owing to the molecular stiffness of incorporated segment. In order to investigate the effectiveness of this synthetic approach, EDC treatment was selected as state-of-the-art reference method, since it has been shown to promote the formation of water-stable collagen materials with no residual toxicity, in contrast to aldehyde biomaterial fixation [1]. Formed hydrogels were investigated for molecular conformation, network architecture and hydrogel macroscopic properties, in comparison to EDC-crosslinked collagen. Furthermore, material degradability and bioactivity were addressed via incubation in SBF, in order to elucidate materials' potential for biomimetic mineralization.

EXPERIMENTAL DETAILS

1,3-Phenylenediacetic acid (Ph) was supplied by VWR International, all the other chemicals were purchased from Sigma-Aldrich. Type I collagen was isolated in-house from rat tail tendons [10]. Collagen was dissolved in 10 mM hydrochloric acid and functionalised with N-(3-Dimethylaminopropyl)-N′-ethylcarbodiimide hydrochloride (EDC)-activated Ph under gentle shaking at room temperature. EDC-crosslinked collagen was synthesized by mixing EDC with collagen solution, as previously reported [5]. Resulting hydrogels were washed with distilled water and dehydrated in aqueous solutions of increasing ethanol concentrations. Attenuated Total Reflectance Fourier Transform Infrared (ATR-FTIR) spectroscopy was carried out on dry samples using a Perkin-Elmer Spectrum BX spotlight spectrophotometer with diamond ATR attachment. 64 scans were averaged for each spectrum, using 4 cm^{-1} resolution and 2 cm^{-1} scanning interval. Degree of crosslinking (C) of collagen networks was determined by 2,4,6-trinitrobenzenesulfonic acid (TNBS) colorimetric assay (n=2) [11], as the molar ratio between functionalised and pristine, non functionalised lysines. Swelling tests (n=3) were carried out by incubating dry samples in 5 mL distilled water for 24 hours. Water-equilibrated samples were retrieved, paper-blotted and weighed. The weight-based swelling ratio (SR) was calculated as $SR=(m_s\text{-}m_d)/m_d \cdot 100$, where m_s and m_d are swollen and dry sample weights, respectively. Hydrogel discs (ø 0.8 cm, n=4) were compressed (Instron 5544 UTM) with a compression rate of 3 mm·min^{-1}. Differential Scanning Calorimetry (DSC) temperature scans were conducted on 10-140 °C temperature range with 10 °C·min^{-1} heating rate (TA Instruments Thermal Analysis 2000 System and 910 Differential Scanning Calorimeter cell base). Mineralization experiment was carried out via 1-week sample incubation (n=4) at 25 °C SBF, with 0.7 sample weight/solution volume ratio [12]. Retrieved samples were rinsed with distilled water, dried, weighed and gold-coated for SEM/EDS (JEOL SM-35) analysis.

DISCUSSION

Molecular organization of crosslinked collagen via ATR-FTIR spectroscopy

Triple helix collagen conformation is normally associated with three main amide bands, i.e. amide I at 1650 cm^{-1}, resulting from the stretching vibrations of peptide C=O groups; amide II absorbance at 1550 cm^{-1}, deriving from N–H bending and C–N stretching vibrations; and amide III band centered at 1240 cm^{-1}, assigned to the C–N stretching and N–H bending vibrations from amide linkages, as well as wagging vibrations of CH$_2$ groups in the glycine backbone and proline side chains [13]. Figure 1 indicates that the positions of these amide bands are maintained in Ph-as well as EDC-crosslinked networks. Furthermore, the FTIR absorption ratio of amide III to 1450 cm^{-1} band was determined to be close to unity ($A_{III}/A_{1450} \sim 1.01\text{-}1.14$) among the three samples, suggesting preserved integrity of triple helices [14] following functionalisation of native collagen.

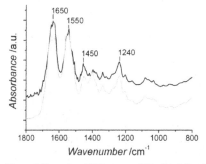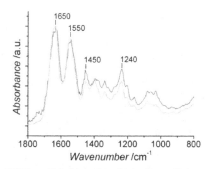

Figure 1. Exemplary FTIR spectra of Ph- (black line, left) and EDC-crosslinked (gray line, right) collagen. Native collagen (light gray line) spectrum is displayed for comparison in both plots.

Network architecture and macroscopic properties of collagen hydrogels

Network architecture was investigated both by quantifying the degree of crosslinking (C) via TNBS assay and by assessing the swelling ratio of resulting materials. As observed in Table I, collagen-Ph samples displayed a higher degree of crosslinking compared to state-of-the-art EDC-crosslinked collagen, despite having a very low Ph-collagen lysine molar ratio.

Table I. Degree of crosslinking (C), swelling ratio (SR) and denaturation temperature (T_d) of collagen hydrogels. *Sample are coded as 'Collagen-XXX-YY', where XXX indicates the type of system (either Ph or EDC-based), while YY identifies the molar ratio of either Ph carboxylic functions or EDC to collagen lysines.

Sample ID*	C /mol.-%	SR /wt.-%	T_d /°C
Collagen-Ph0.5	88 ± 3	1285 ± 450	80
Collagen-Ph1	87 ± 14	1311 ± 757	80
Collagen-Ph1.5	> 99	823 ± 140	88
Collagen-EDC10	25 ± 6	1392 ± 82	68
Collagen-EDC20	37 ± 13	1595 ± 374	76
Collagen-EDC30	34 ± 1	1373 ± 81	78
Collagen-EDC40	68 ± 3	1374 ± 182	80
Collagen-EDC60	60 ± 1	1106 ± 82	80

A slight increase of Ph content (0.5 to 1.5 [COOH]/[Lys] ratio) in the crosslinking mixture led to nearly-complete functionalisation of collagen lysines (> 99 mol.-%). These results suggest that the employment of a bifunctional segment is likely to promote an increased yield of collagen functionalisation, since collagen molecules separated by a distance can be bridged [8]. On the other hand, EDC-mediated functionalisation results in the formation of zero-length net-points, whereby intramolecular crosslinks are likely established [5,8]. Consequently, steric hindrance effects almost certainly explain the decreased degree of crosslinking observed in collagen-EDC, compared to collagen-Ph, samples. Besides the degree of crosslinking, hydrogel swelling behaviour was also investigated; the swelling ratio of one sample, collagen-Ph1.5, was found to be significantly lower compared to all other EDC-based samples. This is supported by TNBS results, suggesting nearly complete functionalisation of collagen lysines for this composition. As for the other Ph-crosslinked samples (collagen-Ph0.5/1), swelling ratios were similar to each other, which is again supported by TNBS results, describing a similar degree of crosslinking.

In order to investigate whether variation of molecular parameters could induce changes in macroscopic properties, the thermo-mechanical properties of collagen hydrogels were addressed. Collagen denaturation temperature (T_d) is related to the unfolding of collagen triple helices into randomly-coiled chains; it is therefore expected to be highly affected by the formation of a covalent network [2]. Table I describes hydrogel T_d values as obtained by DSC; crosslinked samples show a denaturation temperature in the range of 68–88 °C, which is found to be higher than the denaturation temperature of native collagen ($T_d \sim 67$ °C). Furthermore, variation of T_d seems to be directly related to changes of crosslinking degree in the hydrogel network. These results give supporting evidence that covalent net-points were established during hydrogel formation, so that collagen triple helices were successfully retained and stabilized. It should be noted that Ph-crosslinked collagen revealed higher denaturation temperatures with respect to EDC- [6], glutaraldehyde- [7], and hexamethylene diisocyanate- [9] crosslinked collagen materials. This indicates that the incorporation of Ph as a stiff, aromatic segment superiorly stabilizes collagen molecules in comparison with current crosslinking methods.

Besides thermal analysis, mechanical properties of collagen-Ph hydrogels were measured by compression tests. Samples described J-shaped stress-compression curves (data not shown), similar to the case of native tissues. Here, shape recovery was observed following load removal up to nearly 50% compression, suggesting that the established covalent network successfully resulted in the formation of an elastic material, as observed in linear biopolymer networks [15]. On the other hand, EDC-crosslinked collagen showed minimal mechanical properties, whereby sample break was observed even after sample punching. Consequently, quantitative data of mechanical properties on collagen-EDC samples could not be acquired. These findings give further evidence that the collagen functionalisation with Ph is effective for the formation of collagen materials with enhanced mechanical properties. Compressive modulus (E: $28\pm10 \rightarrow 35\pm9$ kPa) and maximal stress of Ph-crosslinked samples (σ_{max}: $6\pm2 \rightarrow 8\pm4$ kPa) were measured in the kPa range, while compression at break (ε_b: $53\pm5–58\pm5$ %) did not exceed 60% compression (Figure 2). The resulting compressive modulus was therefore measured to be almost 20 times higher compared to previously-reported collagen-based materials [8]. At the same time, there was little variation in mechanical properties among the different compositions. Given the unique organization of collagen, the hierarchical level at which covalent crosslinks are introduced is crucial in order to study the influence of crosslinking on the mechanical properties of collagen. Olde Damink et al. observed no variation of mechanical properties in dermal sheep crosslinked collagen [6,7,9]. This was explained based on the fact that crosslinks were mainly introduced within rather than among collagen molecules. This hypothesis may be supported by above mechanical findings, although it is not in line with TNBS, swelling and thermal analysis data. Most likely, the variation of Ph feed ratio among the different compositions ($0.5 \rightarrow 1.5$ [COOH]/[Lys] ratio) was probably too low to result in significant changes in mechanical properties. For these reasons, a wider range of Ph concentrations may be advantageous in order to establish hydrogels with varied mechanical properties. In that case, investigation via AFM will be crucial in order to explore the hierarchical levels at which covalent crosslinks are introduced.

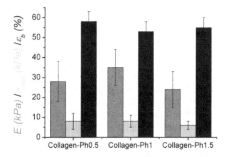

Figure 2. Compressive modulus (E), maximal compressive stress (σ_{max}), and compression at break (ε_b) of Ph-crosslinked collagen hydrogels (n=4).

SBF incubation of collagen hydrogels

SBF incubation is a well-known method to test a material's ability to form a hydroxycarbonate apatite (HCA) layer *in vitro* [12]. Collagen is known to trigger bone-like apatite deposition *in vivo* during bone formation [4], so it was of interest to investigate collagen hydrogel behaviour in SBF as an osteogenic-like medium. The mass change as well as the presence of calcium/phosphorous elements was therefore quantified in retrieved samples in order to (i) clarify any occurrence of degradation and (ii) determine the chemical composition of any potentially-nucleated phases.

A slight decrease in mass (averaged mass loss ~ 4 wt.-%) was observed in collagen-Ph samples, indicating minimal hydrolytic degradation had occurred. SEM on retrieved samples revealed nearly-intact material surfaces, confirming that a covalent network was still present at the molecular level. Among the different compositions, only one sample, collagen-Ph1, displayed a slight mass increase; observed differences in hydrolytic degradation may be related to a varied crosslinking/grafting ratio in the formed hydrogel networks. Networks with increased yield of grafting will likely degrade faster compared to networks with increased yield of crosslinking, since grafted molecules are expected to be cleaved more easily by water compared to crosslinked molecules. Consequently, variation of Ph feed ratio may affect the yield of crosslinking, so that grafted as well as crosslinked molecules may be formed above a specific Ph ratio threshold. A much higher mass loss (averaged mass loss ~ 53 wt.-%) was observed in EDC-compared to Ph-crosslinked collagen, which is in line with previous findings. Here, small micro-pores were observed on the retrieved sample surface, suggesting a surface, rather than bulk, erosion mechanism of hydrolytic degradation.

Besides degradation behaviour, SEM/EDS analyses were carried out to explore whether any mineral phase was nucleated following SBF incubation. Here, sample washing with water was crucial in order to remove superficial deposition of magnesium, sodium and chlorine ions, in agreement with the use of SBF. The presence of calcium and phosphorous elements was observed in all samples, although with a low Ca/P atomic ratio (0.84-1.41). This suggests that the mineral phase laid down on the material surface was most likely constituted of amorphous calcium phosphate, which may be expected due to the relatively short incubation time (1 week) at room instead of body temperature. Interestingly, sample collagen-Ph1.5 displayed increased

Ca/P atomic ratio (Ca/P ~ 1.41) compared to the other samples, likely hinting at enhanced and selective nucleation of an apatite layer in hydrogels crosslinked with increased Ph feed ratio.

CONCLUSIONS

This study highlights the important role played by network molecular architecture on the macroscopic properties and functions of resulting collagen hydrogels. Following a bottom-up synthetic approach, functionalisation with a bifunctional, aromatic segment, successfully led to the establishment of a biomimetic system with preserved triple helix integrity and enhanced macroscopic properties, in comparison to state-of-the-art crosslinked collagen. Resulting hydrogels displayed minimal hydrolytic degradation following 1-week incubation in SBF, whereby nucleation of amorphous calcium phosphate phase was initiated.

ACKNOWLEDGEMENTS

This work was funded through WELMEC, a Centre of Excellence in Medical Engineering funded by the Wellcome Trust and EPSRC, under grant number WT 088908/Z/09/Z. The authors would like to thank J. Hudson, W. Vickers and S. Finlay, for kind assistance with SEM/EDS, SBF preparation, and compression tests, respectively.

REFERENCES

[1] T. Huynh, G. Abraham, J. Murray, K. Brockbank, P.-O. Hagen, S. Sullivan, *Nature Biotechnology* 17, 1084 (1999).
[2] L. Meng, O. Arnoult, M. Smith and Gary E. Wnek, *J. Mater. Chem.* 22, 19414 (2012).
[3] N. Davidenko, T. Gibb, C. Schuster, S.M. Best, J.J. Campbell, C.J. Watson, R.E. Cameron, *Acta Biomater.* 8, 667 (2012).
[4] Y. Wang, T. Azaïs, M. Robin, A. Vallée, C. Catania, P. Legriel, G. Pehau-Arnaudet, F. Babonneau, M.-M. Giraud-Guille, and N. Nassif, *Nature Mater.* 11, 724 (2012).
[5] S. Yunoki and T. Matsuda, *Biomacromolecules* 9, 880 (2008).
[6] L.H.H. Olde Damink, P.J. Dijkstra, M.J.A. van Luyn, P.B. van Wachem, P. Nieuwenhuis and J. Feijen, *Biomaterials* 17, 772 (1996).
[7] L.H.H. Olde Damink, P.J. Dijkstra, M.J.A. Van Luyn, P.B. Van Wachem, P. Nieuwenhuis and J. Feijen, *J. Mater. Sci. Mater. Med.* 6, 465 (1995).
[8] M.G. Haugh, C.M. Murphy, R.C. McKiernan, C. Altenbuchner, and F.J. O'Brien, *Tissue Eng A Part A* 17, 1202 (2011).
[9] L.H.H. Olde Damink, P.J. Dijkstra, M.J.A. Van Luyn, P.B. Van Wachem, P. Nieuwenhuis and J. Feijen, *J. Mater. Sci. Mater. Med.* 6, 431 (1995).
[10] E. Bell, B. Ivarsson, and C. Merrill, *Proc. Natl. Acad. Sci.* 76, 1274 (1979).
[11] W.A. Bubnis and C.M. Ofner, *Analyt. Biochem.* 207, 129 (1992).
[12] S.K. Misra, T. Ansari, D. Mohn, S.P. Valappil, T.J. Brunner, W.J. Stark, I. Roy, J.C. Knowles, P.D. Sibbons, E.V. Jones, A.R. Boccaccini and V. Salih, *J. R. Soc. Interface* 7, 454 (2010).
[13] B.B. Doyle, E.G. Bendit, and E.R. Blout, *Biopolymers* 14, 940-944 (1975).
[14] L. He, C. Mu, J. Shi, Q. Zhang, B. Shi, W. Lin, *Int. J. Biol. Macromol.* 48, 356 (2011).
[15] G. Tronci, A.T. Neffe, B.F. Pierce, A. Lendlein, *J. Mater. Chem.* 20, 8881 (2010).

Mater. Res. Soc. Symp. Proc. Vol. 1498 © 2012 Materials Research Society
DOI: 10.1557/opl.2012.1560

Demonstration of Molecular Sensing Using QCM Device Coated with Stimuli-sensitive Hydrogel

Yoshimi. Seida[1,2], Yuri, Nakano[2] and Yoshio Nakano [2]
[1]Natural Science Laboratory, Toyo University, Tokyo, 112-8606 Japan
[2]Interdisciplinary Graduate School of Science and Technology,
Tokyo Institute of Technology, Yokohama, 226-8504 Japan

ABSTRACT

Molecular sensing using stimuli-responsive viscoelastic property of amphiphilic polymer hydrogel was demonstrated. Thermo-responsive poly(N-isopropylacrylamide); NIPA, hydrogel immobilizing bovine serum albumin (BSA) in its polymer chain was synthesized on the AT-cut QCM (Quartz crystal microbalance). The device was provided to the sensing of anti-BSA antibody. The resonance behavior of QCM in response to the adsorption of anti-BSA antibody on the hydrogel was observed based on both the resonance frequency and resistance of QCM with impedance analysis. The QCM device coated with the LCST polymer hydrogel revealed a characteristic resonance behavior in response to the adsorption of anti-BSA antibody in the collapse phase of gel at which the gel lost most of its hydrated water. The resonance frequency and resistance in the collapsed gel were highly sensitive and their changes in response to the target adsorption were much larger than in the swelling state of gel. The use of the device at the LCST of NIPA gel is ideal because the largest amplification of the adsorption signal is available along with the largest phase change of NIPA gel.

INTRODUCTION

The phase change of stimuli-responsive hydrogels can be interpreted to occur through a change in hydration of the constituent polymer. The change in hydration is able to be induced by a change in surrounding conditions or by some stimuli on the gels. According to the principle of phase behavior in the hydrogels, the phase change of hydrogel in response to adsorption of foreign molecules is also available. An adsorption that alters the hydration of polymer will induce a large change of property of gel. Based on this simple assumption, the gels will be candidate of molecular sensor that is working via the adsorption-induced phase change of gel. The phase change of the gels is able to monitor by quartz crystal microbalance with high sensitivity and resolution [1]. Viscoelastic change of the hydrogel occurring along with the change in the hydration of hydrogels is a good measure of gel in the application as a sensor material. The simple sensing of viscoelasticity of gel can be performed through coupling the gels with QCM [1,2].

In the present study, molecular sensing via adsorption-induced phase change of gel was examined using antigen-antibody reaction. Antigen-immobilized N-isopropyl acrylamide; NIPA gel was synthesized and coupled with QCM to demonstrate the molecular sensing based on the viscoelastic phase property of the stimuli-responsive hydrogel in response to the target adsorption.

EXPERIMENT

Synthesis of BSA-Immobilized NIPA Gel

Regent grade N-hydroxy-succinimide ester, albumin from bovine serum Cohn Fraction V, pH 7.0 (BSA; Wako Co., biochemical agent), N,N'-methylenebisacrylamide (BIS; crosslinker), N,N,N',N'-tetramethylethylene diamine (TEMED; accelerator), ammonium persulfate (AP; initiator) and phosphate buffered saline; PBS solution pH 7.4 (0.02 mol/L) were used as received. N-isopropylacrylamide (NIPA) was supplied from Kohjin Co.Ltd. Japan, and was used after recrystallization from acetone/hexanes (4:6 v/v) mixture. Anti-albumin, bovine serum (Anti-BSA antibody) was supplied from Inter-cell technologies, Inc..

In order to immobilize BSA in polymer network of gel, modification of BSA with vinyl group was performed using N-succinimidyl acrylate (NSA) according to the method reported elsewhere [3,4]. 10 ml phosphate buffer solution (PBS) containing 10 mg/ml BSA was equilibrated at 309 K. A fresh 44.8 mM of NSA dissolved in N,N-dimethylformamide solution was mixed with the BSA solution with NSA/BSA molar ratio 60:1. The mixture was incubated at 309 K for 1 hour to obtain the BSA modified by vinyl group. The obtained modified-BSA was purified by means of gel filtration using Econo-Pac 10DG Columns.

Reaction mixture of hydrogel was prepared by mixing 1 mL of PBS solution containing NIPA (0.5 M), BIS (10.9 mM), TEMED (21.6 mM), AP (17.5 mM), and 500 μL of the modified BSA solution. 0.6 μL of the reaction mixture was dropped on the surface of 5MHz AT-cut quartz resonator to produce thin hydrogel. The thin micro-volume hydrogel prepared on the resonator was rinsed well with PBS for one day. Introduction of BSA in polymer chain of the gel was confirmed by UV-VIS spectrophotometory based on the tyrosine and tryptophan absorption (Abs = 280 nm) in the BSA [6]. The absorption spectra of both BSA-immobilized NIPA gel and pure NIPA gel (the NIPA gel without BSA) were measured (190 ~ 400 nm) to confirm the immobilization of BSA in the gel.

QCM Measurement

The QCMs coupled with the NIPA gel and the BSA-immobilized NIPA gel were immersed in 10ml PBS (pH7.4) solutions containing anti-BSA antibody (6.3×10^{-6}, 6.3×10^{-5} and 6.3×10^{-4} mM), respectively. They were incubated at 309 K for 1 h to complete antigen-antibody reaction first. Then, the temperature of each sample was controlled (295 ~ 333 K) followed by the QCM measurement at each temperature. The resonance frequency shift, $-\Delta F$, and the resonance resistance, ΔR before and after the adsorption of antibody were measured using impedance analyzer. The $-\Delta F$ is the measure of mass and/or elasticity of load on the QCM. The R is the measure of viscosity of load. The frequency shifts and the shift of resonance resistance ΔR with reference to bare QCM under the same conditions were used for each evaluation.

RESULTS AND DISCUSSION

Viscoelastic Behavior of BSA-Immobilized NIPA Gel

The $-\Delta F$ and ΔR of the gels in PBS were much smaller than those in pure water, especially above the LCST region due to a decrease of osmotic pressure working on the gel in PBS [7]. The shift of $-\Delta F$ and ΔR toward higher temperature region was observed and the LCST as well. The shift of LCST will be due to the increase of hydrophilicity of NIPA gel by the introduction of BSA in the network. Figures 1 (a) and (b) show temperature dependence of the resonance frequency shift $-\Delta F$ and the resonance resistance ΔR of the NIPA gel and BSA-immobilized NIPA gel in PBS, respectively. The immobilization of BSA in the NIPA network affects the viscoelastic behavior of NIPA gel and its lower critical solution temperature (LCST) significantly. Figure 1(c) showed the $-\Delta F$ vs. ΔR plots obtained from the data shown in Figs. 1(a) and (b). The $-\Delta F$ vs. ΔR pattern of the BSA-immobilized NIPA hydrogel is larger than the pattern of pure NIPA gel, indicating the immobilization of BSA in the network of gel. The introduction of the BSA modified the gel more hydrophilic and viscoelastic in its collapse phase above LCST.

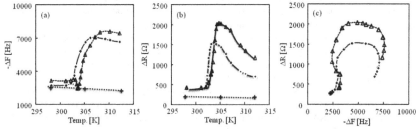

Figure 1. Temperature dependence of (a) the resonance frequency shift ($-\Delta F$) and (b) the resonance resistance (ΔR) of the NIPA gel ($\bullet$), the BSA-immobilized NIPA gel (Δ) and the bare QCM (+). (c) $-\Delta F$ vs. ΔR plot.

QCM response of the BSA-Immobilized NIPA Gel in Contact with Antibody

Figure 2 shows the time course of the $-\Delta F$ and the ΔR during adsorption of anti-BSA antibody onto the BSA-immobilized NIPA gel at 297.8 K (initial concentration of the anti-BSA antibody $C_0 = 6.3 \times 10^{-4}$ mM). The $-\Delta F$ increased quickly just after injection of the antibody, indicating the adsorption of antibody onto the BSA in the gel. On the other hand, little change of resonance resistance R (ΔR) was observed. Therefore, it was considered that the influence of the viscosity change on $-\Delta F$ is small at the temperature [8,9]. The amount of adsorbed antibody was calculated based on the Sauerbrey's equation [10]. 178 ng antibody adsorbed on the BSA-immobilized NIPA gel.

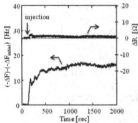

Figure 2. Time-course of the resonance frequency (-ΔF) and the resonance resistance (ΔR) during the adsorption of the antibody at 297.8K

Figures 3(a) and (b) show the temperature dependence of the −ΔF and the ΔR in the BSA-immobilized NIPA gel before and after the adsorption of anti-BSA antibody. Above the LCST of the gel, a large increase of the resonance frequency shift (−ΔF) with a remarkable decrease of the resonant resistance (ΔR) was observed after the adsorption of anti-BSA antibody. The QCM behaviors of pure NIPA gel under the same conditions are shown in Figures 4(a) and (b). Little change in the −ΔF and ΔR was observed in the case of pure NIPA gel without immobilized BSA. The pure NIPA gel will slightly interact with the antibody as can be seen in the QCM responses in Figs. 4(a) and (b). Fig.4(c) indicates the −ΔF vs. ΔR plot. The −ΔF vs. ΔR curve shown in Fig. 4(c) indicates a slight shift of the curve toward smaller resonance resistance (ΔR). The viscoelastic behavior of BSA-immobilized NIPA gel changed obviously due to the adsorption of antibody on the BSA-immobilized NIPA gel. The relative viscosity per loading mass of the collapsed gel decreases with increasing cross-linking of polymer in case of homogeneous hydrogel [11]. The relative viscosity of BSA-immobilized gel after adsorption the antibody will decrease due to multiple-points binding between the anti-BSA antibody and the immobilized antigen (BSA) in the gel network [4]. The results in figure 3 correspond to this assumption.

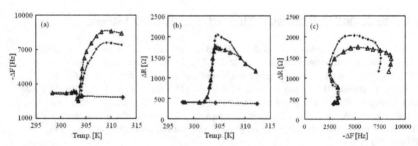

Figure 3. Temperature dependence of (a) the resonant frequency shift (−ΔF) and (b) the resonant resistance (ΔR) of the BSA-immobilized NIPA gel. Before (♦), after (Δ) the adsorption of antibody and the bare QCM (+). (c) −ΔF vs.ΔR plot.

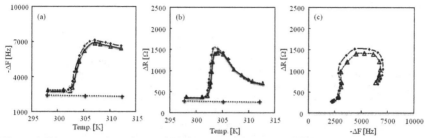

Figure 4. Temperature dependence of (a) the resonant frequency shift ($-\Delta F$) and (b) the resonant resistance (ΔR) of pure NIPA gel before and after the antibody adsorption (C_0=6.3x10^{-4} mM) and the bare QCM (+). (c) $-\Delta F$ vs. ΔR plot

The sensitivity of the BSA-immobilized NIPA gel in response to the adsorption of anti-BSA antibody was elucidated for the series of antibody concentration. Figure 5(a) indicates the resonant frequency shift of the QCM in response to the adsorption of anti-BSA antibody at each temperature for the series of antibody concentration. The gel in its collapse phase (> LCST) showed the shift of resonant frequency much larger than its swelling phase (< LCST). The largest shift was induced at LCST of the gel. Figure 5(b) indicates the adsorption-induced frequency shift at 298 (< LCST), 304 (~ LCST) and 310 K (> LCST). Higher sensitivity of the gel above LCST is also obvious from this figure.

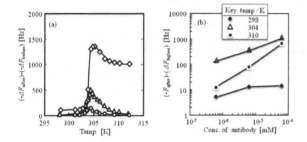

Figure 5. (a) Temperature dependence of the frequency shift ($-\Delta F$) due to the adsorption of the antibody (from 6.3x10^{-4} mM solution) in the BSA-immobilized NIPA (Initial concentrations of the anti-BSA antibody are 6.3x10^{-4} (◇), 6.3x10^{-5} (▲) and 6.3x10^{-6} mM (○), respectively). (b) The adsorption-induced frequency shift at 298, 304 and 310 K as a function of anti-BSA antibody concentration.

155

CONCLUSIONS

The bovine serum albumin (BSA) was covalently immobilized on the network of NIPA hydrogel to produce the hydrogel that is sensitive to anti-BSA antibody. Molecular sensing based on the viscoelastic change of stimuli-responsive hydrogel (immobilizing antigen-antibody reaction site) was demonstrated via coupling the gel with QCM. The viscoelastic behavior of BSA-immobilized NIPA gel in the presence of antibody of BSA was observed as a function of temperature and the concentration of anti-body. The change in viscoelasticity of the BSA-immobilized NIPA gel in contact with the anti-BSA antibody was observed by the QCM with high resolution and sensitivity. The resonance frequency and resistance in the collapsed gel were highly sensitive and their changes in response to the target adsorption were much larger than in the swelling state of gel. The use of the device near the LCST of NIPA gel is ideal because the largest increase of adsorption signal is available due to the large phase change of NIPA gel at LCST. The feasibility of the stimuli-responsive hydrogel coupled with the QCM was confirmed.

REFERENCES

1. Y. Seida, I. Sato, K. Taki and Y. Nakano, *Trans. Mat. Res. Soc.*, **32**, 783 (2007)
2. Paul, S., D. paul, T. Basova, and A. K. Ray, *J. Phys. Chem. C*, **112**, 11822(2008)
3. Shoemaker, S. G., A. S. Hoffman, and J. H. Priest, *Applied Biochemistry and Biotechnology*, **15**, 11(1987)
4. Miyata, T., N. Asami, and T. Uragami, *Nature*, **399**, 766 (1999 a)
5. Lu, Z. R., P. Kopeckova, and J. Kopecek, *Macromol. Biosci.*, **3**, 296(2003)
6. Aitkan, A. and M. P. Learmonth; "Protein Determination by UV Absorption," The Protein Protocols Hand Book, 2nd Edition, Humana Press Inc., Totowa, NJ, 3-5 (2002)
7. Seida, Y. and Y. Nakano, *J. Chem. Eng. Japan*, **29**, 767 (1996)
8. Kanazawa, K. and J. G. Gordon, *Anal. Chim. Acta*, **175**, 99 (1985)
9. Seida, Y., *Proc. 10th Japan-Korea Symposium on Materials & Interfaces, Proc. Int. Symposium on Frontiers in Chemical Engineering*, Kyoto, 16(2012)
10. Sauerbrey, G., *Z. Für. Phys.*, **155**, 206 (1959)
11. Y. Nakano, Y. Seida and Y. Nakano, *J. Chem. Eng. Japan*, **42**(7), 531(2009)

Bioinspired Directional Surfaces–From Nature to Engineered Textured Surfaces

Mater. Res. Soc. Symp. Proc. Vol. 1498 © 2013 Materials Research Society
DOI: 10.1557/opl.2013.105

Structures and Function of Remora Adhesion

Jason H. Nadler[1], Allison J. Mercer[1], Michael Culler[2], Keri A. Ledford[1], Ryan Bloomquist[3], and Angela Lin[2,4]

[1]Georgia Tech Research Institute, Atlanta, GA 30332, U.S.A.

[2]Woodruff School of Mechanical Engineering, Georgia Institute of Technology, Atlanta, GA 30332, U.S.A.

[3]School of Biology, Georgia Institute of Technology, Atlanta, GA 30332, U.S.A.

[4]Parker H. Petit Institute for Bioengineering and Bioscience, Georgia Institute of Technology, Atlanta, GA 30332 U.S.A.

ABSTRACT

Remoras (echeneid fish) reversibly attach and detach to marine hosts, almost instantaneously, to "hitchhike" and feed. The adhesion mechanisms that they use are remarkably insensitive to substrate topology and quite different from the latching and suction cup-based systems associated with other species at similar length scales. Remora adhesion is also anisotropic; drag forces induced by the swimming host increase adhesive strength, while rapid detachment occurs when the remora reverses this shear load. In this work, an investigation of the adhesive system's functional morphology and tissue properties was carried out initially through dissection and x-ray microtomographic analyses. Resulting finite element models of these components have provided new insights into the adaptive, hierarchical nature of the mechanisms and a path toward a wide range of engineering applications.

INTRODUCTION

Animals have been increasingly studied for their adhesive properties due to the successes of fabricated bio-inspired adhesives [1, 2]. Adhesive systems have been studied in geckos, tree frogs and insects [3], which have the common characteristics of hierarchical organization over multiple size scales and the exploitation of redundant anisotropic structure [4]. Biological adhesive systems exhibit surface features over size scales spanning nanometers to several microns, and adhere with van der Waals forces or capillary forces [1, 3]; however, these forces may be marginalized in a fluid environment [3, 5].

Thus far, the remora (family Echeneidae) (Figure 1) has been overlooked as a potential source of bio-inspired adhesion. The remora has an evolved adhesive dorsal fin [6]: a disk-like pad spanning the majority of its dorsal-anterior surface [7]. This highly modified adaptation allows the remora to attach to a wide variety of marine hosts including sharks, sea-turtles, whales and man-made vessels in order to "hitchhike". Remora adhesion is anisotropic, in that the adhesion is enhanced by the maintenance of a drag force, or a posteriorly-directed shear load on the fish. If the shear load is reversed, the remora is capable of rapid detachment [8]. It has been suggested that remora adhesion is primarily suction-based, however, it has been found that the force of adhesion is significantly enhanced when attached to sharkskin as compared to a smooth surface [8].

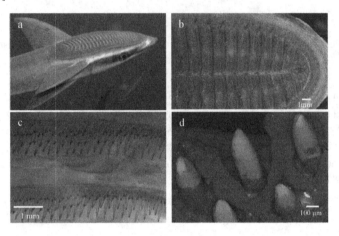

Figure 1. Hierarchical morphology of remora adhesive system. (a) Remora (Echeneis Naucrates, photograph courtesy of Richard Ling). (b,c) Optical micrographs of the (b) Adhesion disk and (c) Pectinated lamellae and spinules. (d) SEM micrograph of tooth-like tips of spinules.

The remora's dorsal disk consists of many developmental modifications to both hard and soft dorsal tissues. The remora's ovoid shaped disk is surrounded by a thick, fleshy lip of connective tissue that completes the external seal to a host [7] [Figure 1(b)]. This lip encloses the highly modified and intricate dorsal skeletal structure that allows for efficient attachment. The disk is filled with successive rows of pectinated lamellae [9]. Arising from each lamella is an array of iteratively patterned, ectodermally derived organs known as spinules [10]. These spinules are well mineralized and resemble the teeth of phylogenetically basal vertebrates such as teleosts [11] and squamates [12] [Figure 1(d)].

The structural morphology of the remora adhesion system and its characteristic size scales were analyzed in order to understand how it adheres, and to what extent the adhesion is active or passive. By imaging the specimen in the attached and detached states using dyed micro-computed tomography (CT), optical microscopy and scanning electron microscopy (SEM), the specimen was digitally reconstructed and characteristic features were able to be measured, such as the dorsal disk aspect ratio, spinule spacing, lamellae length and lamellae angle. The purpose of the measurements was to quantify structural similarities between specimens of significant size difference, and illustrate some applications of the data obtained. All of the measurements were made using the surface model/rendering software Rhinoceros 4.0 on a file in stereolithography format (STL). Statistical analysis of the measurements was made using MATLAB R2009a. An understanding of the fundamental attachment mechanisms was sought with the aim to illuminate a potential alternative adhesive system to other biologically inspired systems.

EXPERIMENTAL

Specimen preparation

All specimens were cared for and euthanized in accordance with the guidelines and principles of the Georgia Institute of Technology Institutional Animal Care and Use Committee, protocol A11085. Remora specimens were euthanized by a 0.5 g/L overdose of MS-222 (tricaine methanesulphonate). Following euthanizing and dissecting, specimens were fixed for microscopy or micro-CT scanning in a phosphate buffer solution of 10 volume percent formalin for a minimum of 24 hours. After fixing, specimens were washed twice and stored in 1X phosphate buffer.

For specimens imaged under SEM, samples underwent solvent dehydration via submersion in increasing concentrations of ethanol in deionized water. Drying by room temperature sublimation was carried out by initially submerging in a solution of 50% hexamethyldisilazane (HMDS) in ethanol, followed by two subsequent submersions in pure HMDS. Specimens were then removed from the HMDS and allowed to dry in ambient conditions. Some specimens were then carbon coated for imaging. SEM images were obtained using a Hitachi S-4700 FEG operating at an accelerating voltage of 0.5keV, and in some cases, a Robinson backscatter detector at a working distance of 30mm.

To provide contrast enhancement of soft tissues in micro-CT scanning, an ionic contrasting agent was introduced. This solution contained ioxaglate, Hexabrix 320 (Covidien, Hazelwood, MO), which provides 6 iodine molecules per anion. Remora sections of interest were submerged in 20mL of 40% Hexabrix/60% ion-free PBS solution for 7 days.

Micro-CT

The remora fin was scanned in air using a uCT50 (Scanco Medical, Bruttisellen, Switzerland). X-ray source and scan settings were $E = 55$kVp, $I = 200$uA, power $= 11$W, integration time $= 800$ms, field of view (FOV) $= 50$mm, isotropic voxel size $= 16$um. Raw data reconstruction to 2D greyscale tomograms was performed automatically with a convolution backprojection algorithm (pixel matrix 3092x3092, mu-scaling 2048). Greyscale tomograms were processed with global segmentation parameters to filter noise and background from tissue and used to generate binary 3D renderings. Tomograms were also converted to dicom format for further image processing in OsiriX v5.5 [13]. At this stage, images were analyzed via surface, volume or multi-planar reconstruction (MPR) rendering.

Dermal Denticle Spacing on Shark Skin

Multiple species of remora have been observed to attach to mako shark hosts [6]. To provide a basis of comparison for remora dorsal disk spinule spacing, image analysis was performed on secondary composite SEM micrographs of a sample of mako shark (*Isurus*) dermal denticles (Figure 7a) using ImageJ v1.47 image processing software [14]. An initial Gaussian blur using a sigma value of 3 pixels was followed by the application of an Otsu segmentation algorithm (Figure 7b). The resulting binary images were filtered to remove features not associated with those desired and analyzed using particle counting while recording each particle's centroid coordinates (Figure 7c). Each set of centroid coordinates were in turn analyzed to generate a list of nearest neighbors, from which a probability density function (PDF) was constructed using the Expectation Maximization Algorithm implemented in MATLAB [15].

Measurements of Dorsal Disk Structures

The dorsal disk aspect ratio is defined as the ratio of the length of the dorsal disk to the width of the disk in the dorsal view. These dimensions are found by applying a bounding box to the surface model and measuring the length and width of the box.

Spinule spacing is defined here as the minimum distance between the tip of a given spinule and the tips of all other spinules from a given set, or equivalently the tip to tip distance between nearest neighboring spinules. This measurement provides comparable distances for analyzing the geometry of spinule set. The algorithm consists of four steps: surface sectioning, centroid identification, centroid grouping and distance measurement.

An array of closely spaced parallel lines is projected onto the surface model, effectively creating a series of intersections as seen in Figure 2(a). Projections that result in an open cross section are omitted. Centroids of the remaining cross sections are found (b) and grouped based on a threshold distance and a minimum group size, ensuring that all the points in a group are associated with an individual spinule (c,d). The endpoint of each group identifies the tip of each spinule. Using the tip locations, the nearest neighbor of a given spinule is identified by comparing its distance to all the others in the set, producing a list of measurements representing the shortest inter spinule distance. From the set of nearest neighbor distances a probability density function (PDF) is constructed using the Expectation Maximization Algorithm implemented in MATLAB [15].

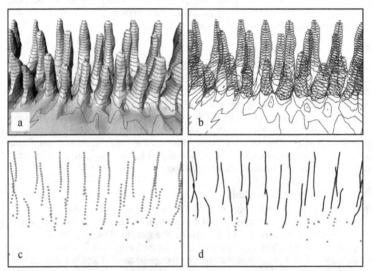

Figure 2: Illustration of the intermediate steps in the spinule spacing algorithm: surface sectioning (a), centroid identification (b), centroid locations (c), and grouping (d).

Lamellae Angle and Length Measurements

Lamellae length is defined here as the length of a line parallel to an individual lamella extending from the midline of the specimen to the perimeter of the sucking disc; L in Figure 3. The lamellae angle measurement is made between the same parallel line, as described by the lamellae length definition, and the specimen midline; θ in Figure 3. The required parallel line for both measurements is obtained by performing a least squares regression of the spinule tip locations.

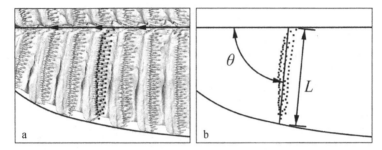

Figure 3: (a) Spinule locations, and specimen midline with arrows indicating posterior to anterior direction; (b) Lamellae Angle (θ) and Length Measurements (L)

RESULTS AND DISCUSSION

Comparative Dorsal Disk Structural Measurements

Micro-CT-based volume renderings of the dorsal disk's outer soft and mineralized tissues are shown in Figure 4. The overall disk dimensions and aspect ratios can be found in Table 1. The average aspect ratio of the three specimens is found to be within ±16 percent of that average. This is especially interesting when noting that Remora B is almost 2.5 times the size of both A and C.

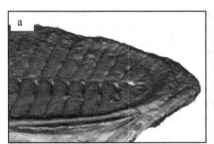
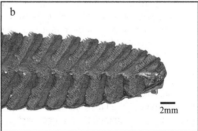

Figure 4: Micro-CT images of suction disk section for segmentation values depicting (a) outer soft tissue and (b) mineralized tissue

Table 1: Remora Disk Aspect Ratio Measurements

Remora Specimen	Length (mm)	Width (mm)	Aspect Ratio
A	38.2	15.7	2.43
B	92.8	28.9	3.21
C	34.7	11.5	3.01
		Average	2.88

Detailed micro-CT-based surface renderings of the pectinated lamellae are shown in Figure 5. The spinule spacing measurements performed on the left half of Remora C, due to the bilateral symmetry of the specimen, can be seen in Figure 7. The histogram has 50 bins each 6.6μm in size. Superimposed on the histogram is a bimodal-normal PDF with parameters described in Table 2.

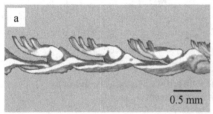 

0.5 mm

Figure 5: Micro-CT rendering pectinated lamellae. (a) 1mm thick MPR sagittal slice at 16 μm resolution, and (b) the spinule detail at (8 μm voxel resolution) of a single lamellae.

The histogram exhibits two distinct peaks, a structure that has been observed by other authors [16]. The first peak can be attributed to a row of shorter, more densely spaced spinules whereas the second peak contains another row of longer, less densely spaced spinules. The PDF confirms that the shorter spinules are more regularly spaced and densely packed than the longer spinules based on the relative height and width of the two peaks. A bimodal-distribution was selected because increasing the fit beyond two produces either extremely low mixing coefficients or clustering of the fitting parameters around the two peak values; fitting three normal distributions to the data shown in Figure 7, for example, gives mixing coefficients of 0.51, 0.48, and 0.01. The spinule spacing on the remora may be compared to the scale or denticle spacing on the host. The likely hood of finding spinules spaced within a given range of values is found by integrating the PDF shown in Figure 7.

Alternatively, the PDF of the host's denticle spacing can be compared directly to that of the remora, as in Figure 7. From the figure, it is apparent that the spinule and denticle spacing are very comparable based on the considerable overlap of the two distributions. These similar length scales may contribute to the larger adhesive force that has been observed when remoras are attached to shark skin compared to smoother surfaces [17].

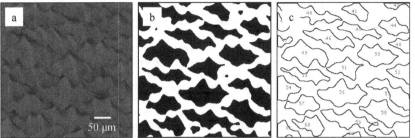

Figure 6: Sequence depicting image processing steps for shark dermal denticle spacing measurements. (a) Original secondary SEM micrograph, (b) Binary image following a blur – segmentation – filter process and (c) Denticle identification and analysis

Table 2: Spinule Probability Density Function Parameters

Parameter	Distribution1	Distribution 2
Type	Gaussian	Gaussian
Mean (μm)	167	116
Std. Dev. (μm)	37.7	12.8
Mixing Coefficient	0.617	0.383
Number of Observations	1821	

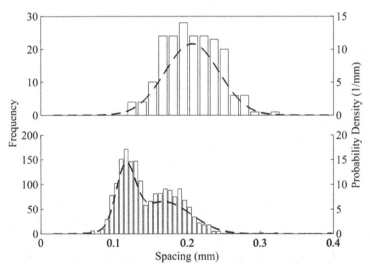

Figure 7: Depiction of shark denticle spacing (top) compared to Remora C spinule spacing (bottom).

Lamellae Length and Angle Measurements

Dissection has revealed that the pectinated lamellae can be elevated or depressed by a series of elegantly arranged musculature to control attachment or release from a host. Lateral depressor muscles originate from the dorsolateral neurocranial vault and attach to the ventrolateral surface of the medial pterygiophore [7]. By serially raising or depressing the lamellae, it was demonstrated that it is likely that remora have evolved an efficient system for a passive attached state of muscular depression. This system permits well-organized "freeloading" until erector activated detachment.

Myoglobin stores oxygen within the muscle tissue of marine animals. Studies have shown concentrations of myoglobin in swimming muscles are significantly higher in those areas that consume the most oxygen during aerobic respiration [18]. For the remora to maintain active muscle control while attached to a marine host over long distances, the muscle tissue surrounding the suction disk would need to have sufficiently high concentrations of myoglobin. Tissue color is a qualitative indicator of myoglobin concentration. During dissection, the remora was found to have very light-colored muscle tissue used for articulating the lamellae and surrounding the suction disk, as compared to the much darker swimming muscles. This is a suggestive indicator for a passive adhesion mechanism. In order to quantitatively measure the concentration and distribution of myoglobin, cell histology will be performed.

Table 3: Lamella Reference Measurements

Remora	Lamella Pairs	Maximum Lamella Length (mm)
A	17	7.90
B	21	14.19
C	23	5.92

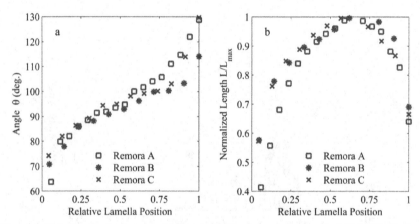

Figure 8: (a) Left half of remora specimens lamellae angle measurements and (b) Left half of remora specimens normalized lamella length measurements

Potential Passive Attachment System

Myoglobin stores oxygen within the muscle tissue of marine animals. Studies have shown concentrations of myoglobin in swimming muscles are significantly higher in those areas that consume the most oxygen during aerobic respiration [18]. For the remora to maintain active muscle control while attached to a marine host over long distances, the muscle tissue surrounding the suction disk would need to have sufficiently high concentrations of myoglobin. Tissue color is a qualitative indicator of myoglobin concentration. During dissection, the remora was found to have very light-colored muscle tissue used for articulating the lamellae and surrounding the suction disk, as compared to the much darker swimming muscles. In order to quantitatively measure the concentration and distribution of myoglobin, cell histology will be performed.

CONCLUSION

Several measurement techniques were described and results were presented for three different remora dorsal suction disks. Similarities in the suction disk structure with respect to the size and position of the lamellae and spinules were illustrated despite significant overall specimen size differences. Remora spinule and shark denticle spacing were found to be of a similar length scale, which may contribute to an increased attachment force between the remora and its host. Lastly, the measurements reveal that further study is needed to fully understand the roles of the different suction disk structural elements and their interactions to create a successful attachment and detachment system.

ACKNOWLEDGMENTS

The authors gratefully acknowledge the support of the Georgia Tech Research Institute, VentureLab, the Georgia Research Alliance, Professor Todd Streelman and Dr. Alistair Dove.

REFERENCES

[1]   K. Autumn, M. Sitti, Y. C. A. Liang, A. M. Peattie, W. R. Hansen, S. Sponberg, T. W. Kenny, R. Fearing, J. N. Israelachvili, and R. J. Full, "Evidence for van der Waals adhesion in gecko setae," *Proceedings of the National Academy of Sciences of the United States of America,* vol. 99, pp. 12252-12256, Sep 17 2002.

[2]   D.-M. Drotlef, L. Stepien, M. Kappl, W. J. P. Barnes, H.-J. Butt, and A. del Campo, "Insights into the Adhesive Mechanisms of Tree Frogs using Artificial Mimics," *Advanced Functional Materials,* pp. n/a-n/a, 2012.

[3]   W. Federle, W. J. P. Barnes, W. Baumgartner, P. Drechsler, and J. M. Smith, "Wet but not slippery: boundary friction in tree frog adhesive toe pads," *Journal of the Royal Society Interface,* vol. 3, pp. 689-697, Oct 22 2006.

[4]   N. A. Malvadkar, M. J. Hancock, K. Sekeroglu, W. J. Dressick, and M. C. Demirel, "An engineered anisotropic nanofilm with unidirectional wetting properties," *Nat Mater,* vol. 9, pp. 1023-8, Dec 2010.

[5]   S. B. Emerson and D. Diehl, "Toe Pad Morphology and Mechanisms of Sticking in Frogs," *Biological Journal of the Linnean Society,* vol. 13, pp. 199-216, 1980.

[6]   B. O'Toole, "Phylogeny of the species of the superfamily Echeneoidea (Perciformes: Carangoidei: Echeneidae, Rachycentridae, and Coryphaenidae), with an interpretation of echeneid hitchhiking behaviour," *Canadian Journal of Zoology*, vol. 80, pp. 596-623, 2002/04/01 2002.

[7]   B. Fulcher and P. Motta, "Suction disk performance of echeneid fishes," *Canadian Journal of Zoology*, vol. 84, pp. 42-50, 2006.

[8]   B. Fulcher and P. Motta, "Suction disk performance of echeneid fishes," vol. 84, p. 42, 2006.

[9]   R. Storms, "X.—The adhesive disk of Echeneis," *The Annals and Magazine of Natural History*, vol. 2, pp. 67-76, 1888.

[10]   E. Gudger, *A Study of the Smallest Shark-suckers (Echeneididae) on Record: With Special Reference to Metamorphosis*: By Order of the Trustees of American Museum of Natural History, 1926.

[11]   G. J. Fraser, R. F. Bloomquist, and J. T. Streelman, "A periodic pattern generator for dental diversity," *BMC Biol*, vol. 6, p. 32, 2008.

[12]   J. M. Richman and G. R. Handrigan, "Reptilian tooth development," *genesis*, vol. 49, pp. 247-260, 2011.

[13]   A. Rosset, L. Spadola, and O. Ratib, "OsiriX: An Open-Source Software for Navigating in Multidimensional DICOM Images," *Journal of Digital Imaging*, vol. 17, pp. 205-216, 2004/09/01 2004.

[14]   J. Schindelin, I. Arganda-Carreras, E. Frise, V. Kaynig, M. Longair, T. Pietzsch, S. Preibisch, C. Rueden, S. Saalfeld, B. Schmid, J.-Y. Tinevez, D. J. White, V. Hartenstein, K. Eliceiri, P. Tomancak, and A. Cardona, "Fiji: an open-source platform for biological-image analysis," *Nat Meth*, vol. 9, pp. 676-682, 2012.

[15]   G. McLachlan and D. Peel, in *General Introduction Finite Mixture Models*, ed: John Wiley & Sons, Inc., 2005.

[16]   R. Britz and G. D. Johnson, "Ontogeny and homology of the skeletal elements that form the sucking disc of remoras (Teleostei, Echeneoidei, Echeneidae)," *Journal of Morphology*, vol. 273, pp. 1353-1366, 2012.

[17]   B. A. Fulcher and P. J. Motta, "Suction disk performance of echeneid fishes," *Canadian Journal of Zoology*, vol. 84, pp. 42-50, 2006/01/01 2006.

[18]   L. K. Polasek and R. W. Davis, "Heterogeneity of myoglobin distribution in the locomotory muscles of five cetacean species," *Journal of Experimental Biology*, vol. 204, pp. 209-15, Jan 2001.

Mater. Res. Soc. Symp. Proc. Vol. 1498 © 2013 Materials Research Society
DOI: 10.1557/opl.2013.332

Encapsulation of Stimuli-Responsive Fusion Proteins in Silica:

Thermally Responsive Metal Ion-Sensitive Hybrid Membranes

Linying Li, [1,3] Owen Im,[1] Ashutosh Chilkoti[1,2,3] and Gabriel P. López[1,2,3]

[1]Department of Biomedical Engineering, Duke University, Durham NC 27708, U.S.A.
[2]Department of Mechanical Engineering and Materials Science, Duke University, Durham NC 27708, U.S.A.
[3]NSF Research Triangle Materials Research Science and Engineering Center, U.S.A

ABSTRACT

In this study, we demonstrate the fabrication of hybrid membranes that exhibit discrete and reversible changes in permeability in response to changes in calcium ion (Ca^{2+}) concentration and temperature. Fusion proteins comprising calmodulin (CAM) and elastin-like polypeptides (ELPs) were used as stimuli-responsive elements due to their ability to undergo a reversible lower critical solution temperature (LCST) phase transition, which is sensitive to Ca^{2+} binding. The calmodulin elastin-like polypeptides fusions (CAM-ELPs) were incorporated into polymerizing silica networks using a simple sol-gel process and spin coating. Permeation experiments with solutions of crystal violet showed that the membranes are both Ca^{2+}-responsive and thermally responsive. Under suitable pressure drop across the membranes, in the absence of Ca^{2+} or below the LCST of the ELPs, the hybrid membranes are impermeable to water. After addition of Ca^{2+} or above the LCSTs, they become permeable to water. The permeability can be toggled back and forth by sequential addition of calcium and ethylenediamine tetraacetic acid (EDTA). These results demonstrate that CAM-ELP/silica hybrid membranes can serve as tunable molecular filters whose permeability can be switched on and off in response to Ca^{2+} and temperature.

INTRODUCTION

Stimuli-responsive materials can exhibit switchable microstructure and properties in response to changes in environmental factors such as temperature, pH, light, magnetic and electric fields, ionic concentration, and the presence of biomolecules. A widely studied class of stimuli-responsive materials are those that experience reversible changes in their microstructure alternating between hydrophilic and hydrophobic states depending on specific stimuli [1]. Such materials have found interest because of their potential for applicability in bioseparations, [2] biosensing systems [3], anti-fouling surfaces [4] and drug delivery systems [5].

Elastin-like polypeptides (ELPs) are a versatile class of stimuli-resonsive polymers and consist of repeated pentapeptide sequences derived from elastin (i.e., Val-Pro-Gly-Xaa-Gly, where Xaa is a "guest residue" that can be any amino acid except proline), and typically exhibit a characteristic lower critical solution temperature (LCST) in water [6]. Below this temperature, these polymers are soluble in water and exist in a hydrophilic, extended conformational state. At the LCST, a reversible phase change occurs such that at higher temperatures the polymers exist in a hydrophobic, collapsed and aggregated structure [7]. Changing the guest residue and molecular weight of ELP sequences allows for adjustment of transition temperatures and thus the thermally responsive character of the polymers. In addition, ELP-fusions have been developed with a number of proteins, including green fluorescent protein [8], thioredoxin [9], and calmodulin (CAM) [10]. By incorporating the responsive ELP domain with proteins that bind ligands (e.g. thioredoxin and CAM), it has been demonstrated that the LCST of the ELP chimera can be modulated and can also be responsive to the binding of ligands.

ELPs fused with the calcium-sensitive protein, CAM, were used in this study; these fusions can undergo conformational changes in response to both ionic (Ca^{2+} binding) and thermal stimuli [11]. Calcium binding with CAM lowers the LCST of the ELP, due to a decrease in the surface charge of the protein or an increase in the hydrophobic surface area [12].

In this study, we show that encapsulation of the CAM-ELP into hybrid silica based membranes allows toggling of membranes between permeable and impermeable states depending on temperature and the presence of ions. Membranes are fabricated using sol-gel processing and spin-coating onto commercial ultrafiltration membranes. The hybrid CAM-ELP silica membrane exhibits molecular switching and becomes impermeable or permeable to aqueous solutions based on the LCST behavior of the ELP. The function of these membranes as switchable filters is demonstrated and tested by determining the permeability of the membranes by visual inspection.

EXPERIMENTAL

2.1. Materials

Tetraethyl orthosilicate (TEOS) (Aldrich), ethyl alcohol (IBI Scientific), crystal violet (MW=408 Da, Fischer Sci.), and ethylenediamine tetraacetic acid (EDTA) (Aldrich) were used as received. All protocols used deionized water (resistivity greater than 18.2 MΩ cm). The CAM-ELPs and ELPs were supplied by the Chilkoti group at Duke University. Synthesis and purification of ELPs were conducted as previously described using an *E. coli* recombinant expression system [13,14]. The ELP composed of 180 pentapeptides in length that has a repeat unit composed of 10 pentapeptides with the guest residues Val, Ala, and Gly in a 5:2:3 ratio, respectively [15]. The CAM-ELPs consisted of a ligand binding protein domain, calmodulin (CAM) and the same ELP domain described above [11].

Using a previously described method [16], a diluted silica sol was prepared with molar ratios between TEOS/ethanol/H_2O/HCl/ surfactant of 1:3:1:0.0007. CAM-ELP (15 mg/ml) solutions were prepared in 1 milliliter aliquots. An aliquot was added to diluted stock sol and stirred to obtain transparent sols, which were then spin-coated at 1500-2000 rpm onto centrifugal filters (Pall Life Sciences NANOSEP MF, active membrane area, 0.3 cm^2) with 30 kDa (NANOSEP 30K) molecular weight cutoffs. For pressurized ultrafiltration experiments, sols were spun onto ultrafiltration disk membranes (Pall Cooperation, 30 kDa molecular weight cutoff, 2.5 cm diameter). Bulk gels were manufactured by curing sols at room temperature for at least 48 hours.

2.2. Permeation experiments

The Ca^{2+} dependent permeability of CAM-ELP membranes to crystal violet solution was measured using the silica/CAM-ELP membranes cured on centrifugal filters. The membranes were immersed sequentially in Ca^{2+} solution and EDTA solution for 30 minutes each. Membrane samples were kept at 32°C ($\pm$ 1 °C) during incubation and centrifugation.

Membranes coated on centrifugal filters were also were tested for temperature-dependent changes in permeability by cycling between 15°C and 40 °C ($\pm$ 1 °C), and centrifuging with crystal violet solution (400 μL, 0.1 wt % aqueous). Temperature equilibration was achieved by incubating samples for 10-20 min before centrifugation. They were maintained at the target temperature during centrifugation (Eppendorf 5424R).

RESULTS AND DISCUSSION

We fabricated silica/CAM-ELP hybrid membranes with different concentrations of CAM-ELPs (6-77 wt%). Permeation tests for Ca^{2+} dependence were conducted over two cycles of Ca^{2+} and EDTA immersion for five samples of each composition (Figure 1, tubes B-F). Figure 1 shows the centrifuged samples with observed filtrate at each addition of solution. For tubes that underwent less than two full cycles, water was added as a control in place of Ca^{2+} or EDTA for subsequent centrifugation steps. Exposure to EDTA, which chelates calcium ions, caused the hybrid membranes to become impermeable to solution. After adding Ca^{2+}, they become permeable. Permeability was toggled on and off by sequential addition of Ca^{2+} and EDTA.

The Ca^{2+}-sensitivities of the membranes were enhanced by increasing the concentrations of CAM-ELPs from 6 to 23 wt%, providing evidence that the Ca^{2+} response is caused by conformational changes of CAM-ELP. The sensitivity remained constant within the range from 23 to 77 wt% ELP. However, the permeability of membranes above 23 wt% ELP was less reversible. As shown in Figure 1, membranes without Ca^{2+} (tubes B and F) also become permeable to water, showing that these membranes are weaker due to low amounts of silica content. Therefore, we conclude that, within the tested sample set, membranes with 23wt% CAM-ELPs concentration have optimal permeability and reversibility characteristics. All further tests were conducted using 23 wt% CAM-ELPs.

Weight ratio (wt %)	Permeation property
6	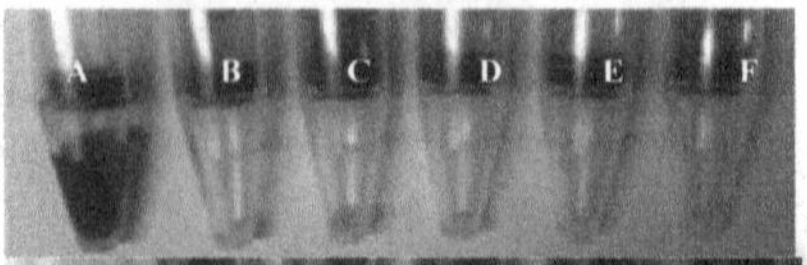
12	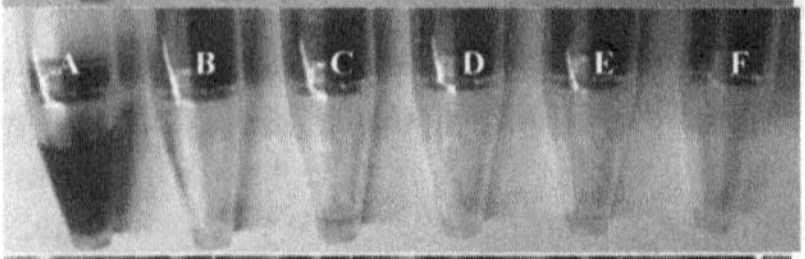
23	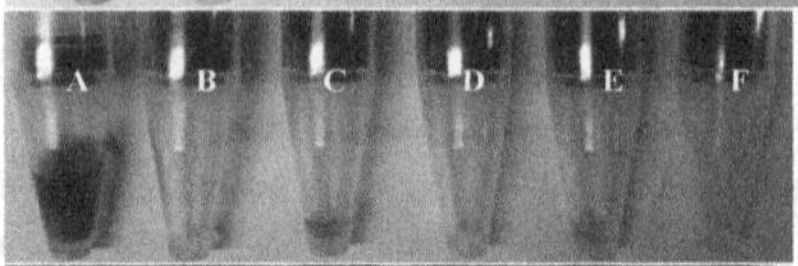
31	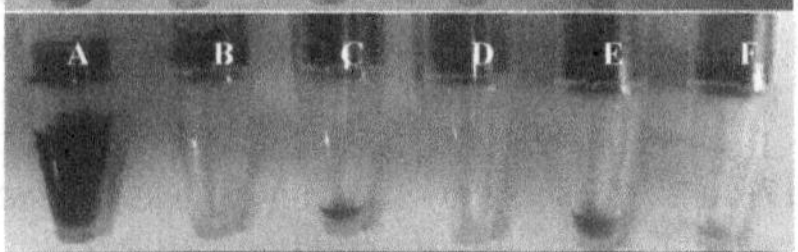
47	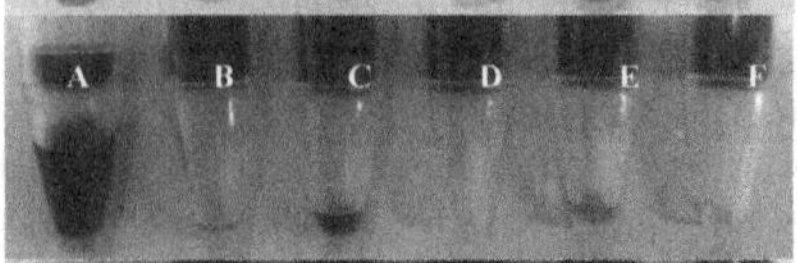
77	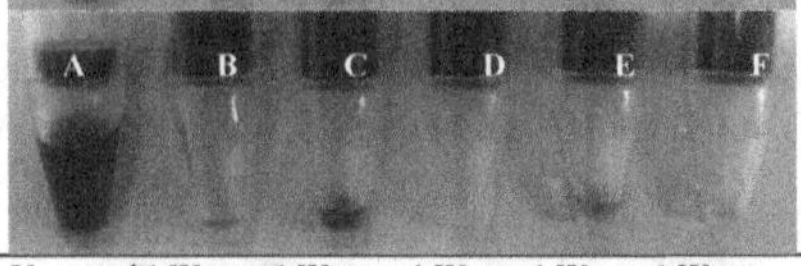
	1. Uncoated 1.Water 1.Water 1.Water 1.Water 1.Water 2.CaCl₂ 2.CaCl₂ 2.CaCl₂ 2.CaCl₂ 3.EDTA 3.EDTA 3.EDTA 4.CaCl₂ 4.CaCl₂ 5.EDTA

Fig 1. Determination of the optimal concentration of ELPs in silica/CAM-ELP hybrid membranes.

Permeation tests for Ca^{2+} dependence were repeated for 23 wt% CAM-ELP hybrid membranes, as well as for ELP / silica membranes without CAM (Figure 2). As before, the CAM-ELP permeability was reversibly toggled on and off by sequential addition of Ca^{2+} and EDTA, and the membranes withstood at least 5 iterations of immersion and centrifugation. We hypothesize that in the absence of Ca^{2+}, the extended conformation of ELPs blocks permeation of the solution. With calcium ions bound to the CAM domain, the ELPs exhibit a contracted conformation that creates a porous network to facilitate permeation through the membrane. Figure 2(b) shows that no crystal violet filtrate was observed for ELP membranes without CAM, indicating that ELP/silica hybrid membranes are not Ca^{2+} dependent. This result further indicates that permeability of the CAM-ELP containing hybrid membranes is controlled by calcium binding-induced LCST transition of CAM-ELPs.

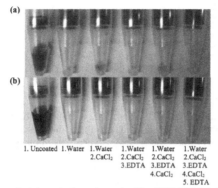

| 1. Uncoated | 1.Water | 1.Water 2.CaCl$_2$ | 1.Water 2.CaCl$_2$ 3.EDTA | 1.Water 2.CaCl$_2$ 3.EDTA 4.CaCl$_2$ | 1.Water 2.CaCl$_2$ 3.EDTA 4.CaCl$_2$ 5. EDTA |

Fig 2. Permeation of crystal violet solutions through silica/CAM-ELP (a) and silica/ELP membranes (b) upon sequential addition of Ca^{2+} followed by EDTA.

CAM-ELP membranes are also temperature sensitive. The reversibility of temperature-induced permeability switches and the stability of the membranes were investigated in tests conducted at 15°C and 40 °C. Figure 2 shows that at 15°C (below the LCST of the CAM-ELPs), the membranes were impermeable to crystal violet solution when centrifuged at 200×g for 3 min. In contrast, when the membranes were centrifuged under similar conditions at 40°C (above the LCST), they became permeable to crystal violet solution. These data strongly suggest that these membranes contain pores that are closed at 15°C and open at 40 °C. In a previous study [17], no permeation was observed for silica/BSA control membranes cycled between 25°C and 40 °C. Therefore, we conclude that the switchable permeability of the membranes is controlled by the LCST behavior of the ELPs. A gradual reduction in centrifugal force was required to retain the reversibility of the membranes (Fig 3). Incomplete reversibility may be due to incomplete closure of pores in the membranes caused by slow relaxation of ELP molecules. A similar trend was observed for the silica/ELP membranes. [17]

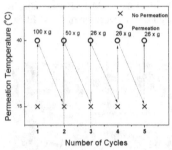

Fig. 3 Permeation experiments of silica/ CAM-ELP hybrid membrane with solutions of crystal violet cycled between 15°C and 40°C. "Permeation" indicates that the solution permeated through the membrane at the indicated centrifugal force. "No permeation" indicates that not even a trace of water was observed to permeate through the filter. Each data point was obtained after 3 min of centrifugation.

CONCLUSIONS

The most salient result of this study was the successful fabrication of responsive CAM-ELP membranes with permeability that can be toggled on and off by temperature modulation and by calcium ion binding. The membranes were shown to undergo a change in conformation due to the LCST behavior of the ELP. The calcium-ion and thermally dependent changes were visually confirmed in permeation experiments, clearly demonstrating that silica/CAM-ELP hybrid membranes can function as switchable molecular filters whose permeability can be controlled by the solubility of CAM-ELP.

ACKNOWLEDGMENTS

This research supported by the NSF Research Triangle MRSEC (DMR-1121107). The NSF is also acknowledged for supporting the travel of Linying Li to the MRS meeting (CMMI-1250333).

REFERENCES

1. P. M. Mendes, *Chem. Soc. Rev.*, **37**, 2512 (2008).
2. M. R. Banki, L. Feng, D.W. Wood, *Nature Methods*, **2**, 659-662 (2005).
3. C. Xu, B. B. Wayland, M. Fryd, K. I. Winey, and R. J. Composto, *Macromolecules*, **39**, 6063 (2006).
4. P. Alexandridis, B. Lindman, *Amphiphilic Block Copolymers: Self-Assembly and Applications*, 1st ed. (Elsevier Science, Netherlands, 2000)
5. A. Chilkoti, M.R. Dreher, D.E. Meyer, D. Raucher, *Advanced Drug Delivery Reviews,* **5**,

613-630 (2002).

6. D. W. Urry, *J. Phys. Chem. B*, **101**, 11007-11028 (1997).

7. D. W. Urry, *Angew. Chem. Int. Ed.*, **32**, 819 (2003).

8. K. T. Carlson, D. E. Meyer, and L. Liu, Peds.**17**,1 (2004).

9. D. E. Meyer, K. T. Carlson, A. Chilkoti, *Biotechnol. Prog* **17**, 720 (2001).

10. K. T. Carlson, L. Liu, B. Kim, A. Chilkoti, *Protein Science*, **13**,12 (2004).

11. B. Kim and A. Chilkoti, *Journal of the American Chemical Society*, **130**, 17867 (2008).

12. S. Banta, I. R. Wheeldon, and M. Blenner, *Annu. Rev. Biomed. Eng.*, **12**, 167 (2010).

13. A. Girotti, J. Reguera, and J. C. Rodríguez-Cabello, *J Mater Sci Mater Med*, **15**, 479 (2004).

14. C. Morrow, D. S. Minehan, J. Wu, and E. Hunter, *Biotechnol Prog*, **8**, 347 (1992).

15. D. E. Meyer and A. Chilkoti A. *Biomacromolecules*, **3**, 357–367 (2002).

16. Q. Fu, G. V. R. Rao, L. K. Ista, Y. Wu, B. P. Andrzejewski, L. A. Sklar, T. L. Ward, and G. P. Lopez, *Advanced Materials*, **15**, 1262 (2003).

17. G. V. R. Rao, S. Balamurugan, D. E. Meyer, A. Chilkoti, and G. P. Lopez, *Langmuir*, **18**, 1819 (2002).

Mater. Res. Soc. Symp. Proc. Vol. 1498 © 2013 Materials Research Society
DOI: 10.1557/opl.2013.333

Triggered Cell Release from Shellac-Cells Composite Microcapsules

Shwan A. Hamad,[1] Simeon D. Stoyanov[2] and Vesselin N. Paunov[1,*]
[1] Department of Chemistry, University of Hull, Hull, HU6 7RX, UK
[2] Unilever R&D, Olivier van Noortlaan 120, 3133 AT Vlaardingen, the Netherlands.

ABSTRACT

We have fabricated novel shellac-cells composite microcapsules capable of pH-stimulus induced release of cells in a narrow pH range. The microcapsules were produced with yeast cells as a model for probiotics which were co-precipitated from an aqueous solution of ammonium shellac doped with pH-sensitive polyelectrolytes. The yeast cells in the composite shellac-cell microcapsules retained their viability even when treated with aqueous solutions of very low pH and subjected to shear stress. We studied the pH triggered release of cells from these microcapsules and measured their disintegration times. These microcapsules showed versatile responses ranging from slow release to explosive swelling at higher pH depending on the type and concentration of the polyelectrolyte integrated in the shellac microcapsules. We also observed growth-triggered release of cells from these microcapsules upon exposure to culture media. In both cases the cells retained their viability following their release from the microcapsules into the aqueous solution.

INTRODUCTION

Shellac is biodegradable and renewable resin of the insect origin [1,2] containing esters of a mixture of polyhydroxy acids [3] with major components being aleuritic and shellolic acids[4-6]. In general, shellac is hydrophobic and possesses good resistance to the gastric fluids which is why it is still used as enteric coating material [7]. It has good solubility in ethanol but it is insoluble in water below pH 7 although it can be dissolved in alkaline solutions [8]. Campbell *et al.* [9-10] showed successful fabrication of food-grade shellac micro-rods with a range of micro-particle inclusions. When yeast cells were used as inclusions, they found that the obtained "lumpy" shellac micro-rods are excellent foam stabilizers which can improve the steric stability of food-grade foams. Law *et al.* recently reported encapsulation using shellac which includes stabilization and targeted delivery of Nattokinase in shellac beads prepared by cross-linking aqueous solution of ammonium shellac salt with calcium ions [2]. Recently a systematic study of the thin-walled liquid-filled pectinate capsules with addition of shellac showed that precipitation of shellac under acidic conditions made the capsules softer and more flexible.[11] Xue *et al.* [12] prepared and characterized calcium-shellac microspheres with added carbamide peroxide as tooth whitening agent. They used two technique based on dropping aqueous shellac solution into calcium chloride solution [12] and alternatively, emulsification of aqueous ammonium shellac solution in sunflower oil with calcium chloride powders [13]. Stummer *et al.* encapsulated probiotics using shellac by coating individual dried yeast cells in a fluid-bed with a mixture of shellac and glycerol, sodium alginate and polyvinylpyrrolidine which produced enteric coating with improved elasticity and solubility and provided protection for the probiotic cells in the intestinal fluid [7]. Here we describe a new technique for producing composite shellac-yeast microcapsules by spray co-precipitating method and introduce pH triggers for slow or accelerated disintegration. The schematics of the microcapsule fabrication are shown in Figure 1.

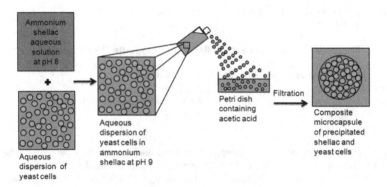

Figure 1: Schematics of the microcapsules preparation method.

EXPERIMENT

Baker's yeast cells were washed multiple times with de-ionized water and re-dispersed in an aqueous ammonium shellac solution. Three different shellac–yeast formulations were prepared: (i) shellac–yeast suspension at pH 9; (ii) shellac–yeast suspension doped with 0.5 wt% carboxymethyl cellulose at pH 9. The pH of these suspensions was adjusted to 9 by 0.1 M sodium hydroxide solution; and (iii) shellac–yeast dispersion doped with 4 wt% sodium polyacrylate at pH 6.7 supported by acetate buffer. The shellac–cell aqueous dispersion was sprayed through a 380 µm metal nozzle with 43 sprays (0.03 g each) per minute over 3 vol% aqueous acetic acid which produced composite "shellac–yeast cell" microcapsules. They were washed with de-ionised water and the viability of the yeast cells inside the microcapsules was tested with Fluorescein diacetate (FDA) by using fluorescence microscopy. The disintegration properties of the composite shellac–yeast microcapsules were studied at various pH corresponding to those in the stomach and the lower intestines at the same conditions upon stirring. The disintegration processes was monitored by withdrawing small aliquots of the media at regular periods and observing them using optical and fluorescence microscopy to count the cells and the remaining microcapsules.

RESULTS AND DISCUSSION

We produced a range of composite shellac-yeast cells microcapsules by spray co-precipitating a dispersion of yeast cells in aqueous ammonium shellac solution. This yielded rigid and stable microcapsules, capable of disintegrating at pH greater than 7. The aqueous droplets of the yeast cell suspension in shellac solution undergo rapid precipitation and partially retain their shape upon their contact with the acetic acid solution. Rigid and stable composite microcapsules were fabricated by spray co-precipitating aqueous dispersion of 10 % wt. yeast cells in 7 % wt.

ammonium shellac solution over 3 % vol. acetic acid solution. The microcapsules were stirred in the acidic medium for 30 minutes then they were filtered off and washed with de-ionized water several times and characterized by optical and fluorescence microscopy as wells as with a SEM (see Figure 2, a-f).

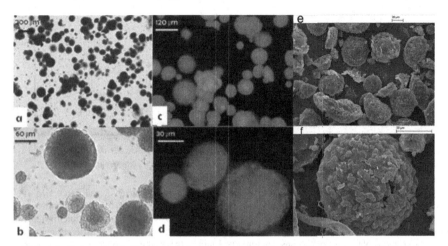

Figure 2: Composite shellac/yeast cells microcapsules fabricated by spray co-precipitating a suspension of 10 % wt. yeast cells in 7 % wt. ammonium shellac solution at pH 8 using 3 % vol. acetic acid. (a,b) Brightfield optical microscopy images; (c, d) Fluorescence microscopy images of the microcapsules where the shellac was doped with Nile Red, (e, f) SEM images of the same samples.

The optimal concentration of ammonium shellac in the initial cell suspension was found to be 7 % wt. which gave sufficient integrity, consistent size range and mechanical stability. We imaged the fabricated microcapsules after doping the ammonium shellac solution with traces of Nile Red (10^{-4} M) before spraying, while the yeast cells inside the fabricated microcapsules were treated with FDA solution to aid their co-localization. We tested the viability of the microencapsulated yeast at pH 1, which is close to the acidity in the stomach. Control samples of unprotected native yeast quickly lost their viability at pH 1 while the microencapsulated cells survived exposure to pH 1 without loss of viability.

The disintegration rate of the microcapsules was controlled by doping the shellac solution with pH-sensitive polyelectrolytes which can swell and subsequently accelerate the microcapsule breakdown to smaller fragments which leads to faster release the cells. Below 0.5 % wt. carboxymethylcellulose (CMC) we obtained stable and rigid composite shellac-yeast microcapsules capable of much faster disintegration and release of the yeast cells at pH 8 as shown in Figure 3. The microcapsules disintegrate and change their morphology and release individual yeast cells in the solution. The graph in Figure 3 shows the percent of remaining microcapsules (relative to the initial count) counted at different time intervals which illustrate the trend of the triggered cell release. Higher concentrations of CMC did not produce microcapsules with sufficient integrity as the CMC interfered with the spray precipitation of the shellac solution at the production stage by increasing the viscosity of the solution.

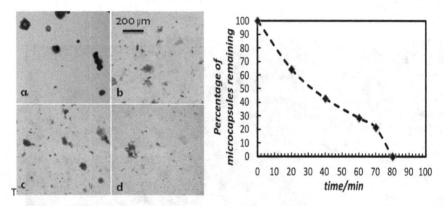

Figure 3: LEFT image: Composite microcapsules of shellac-yeast cells transferred into 0.1 M of phosphate buffer solutions at different pH and stirred on a magnetic stirrer while aliquots were withdrawn at regular intervals and checked by optical microscopy. The above micrographs illustrate the time it takes for the complete disintegration of the microcapsules at different pH. (a) after 24 hours at pH 5, the microcapsules remained intact; (b) is after 7 hours at pH 6.5; (c) is after 3.2 hours at pH 7; (d) is after 2 hours at pH 7.5. RIGHT image: The percentage of remaining composite shellac-yeast microcapsules versus time of incubation (stirring) at pH 8.

Doping of the shellac with sodium polyacrylate (PAA) showed much better results with pH triggered cell release than using CMC. PAA is an anionic polymer which swells in aqueous solutions above pH 7-7.5 and increases its particle size up to 400 times. We doped 5 % wt. aqueous solution of ammonium shellac with 4 % wt. PAA (with respect to the shellac) at pH 6.7 supported by acetate buffer in which the native yeast cells were dispersed and the dispersion was spray co-precipitated over 3 % vol. acetic acid. We tested the swelling properties of the produced composite microcapsules upon exposure to aqueous sodium bicarbonate solution. Figure 4 shows typical swelling behavior of our microcapsules.

CONCLUSIONS

We developed novel composite microcapsules of shellac and yeast cells by spray co-precipitating yeast cell suspension in ammonium shellac solution at pH 8 into acetic acid solution. We found that 7% wt. shellac is the optimal concentration of shellac which gives rigid and stable microcapsules which can disintegrate at pH > 7 for specified time interval and release viable probiotic cells into the solution. We can control the pH trigger for the cell release and the time of microcapsule disintegration by: (a) the shellac concentration and (b) doping the cell suspension with pH-sensitive polymers. We show that microcapsules doped with carboxymethylcellulose and sodium polyacrylate can trigger the cell release at neutral and higher pH and speed up the shellac-cell microcapsule disintegration process. Doping of the shellac solutions with carboxymethylcellulose significantly enhanced the rate of release the yeast cells from the composite microcapsules at neutral and higher pH. The inclusion of sodium polyacrylate in the shellac resulted in much faster disintegration of the microcapsules upon increase of the pH above 7 and in some cases caused instant cell release. We show that the cells have preserved their viability during the microencapsulation and subsequent the release. The composite microcapsules can protect the cells at extremely low pH and still produce viable cells upon disintegration at neutral and higher pH. These composite shellac-cell microcapsules could be used in formulations for protection and delivery of probiotics and other cells with triggered release of the encapsulated cells by using a variety of pH-sensitive food grade polymers. In future developments, such microcapsules could also be used in applications involving triggered release cells in cell implants, including stem cell and live vaccines delivery.

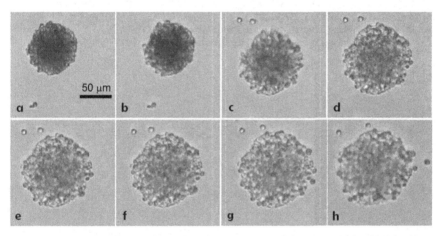

Figure 4: Swelling behavior of the composite shellac/yeast cells microcapsules fabricated by spray co-precipitating yeast cells in ammonium shellac solution doped with 4 % wt. sodium polyacrylate in 3 % vol. acetic acid (aq). The micrographs are time shots the microcapsules swelled at pH 8.

ACKNOWLEDGMENTS
The authors appreciate the EPSRC Industrial CASE award for S. Hamad and the support from EPSRC and Unilever for this research project.

REFERENCES

1.   S.A. Hamad, S.D. Stoyanov, V.N. Paunov, *Soft Matter,* **8**, 50691 (2012).

2.   D. Law, Z. Zhang, *Minerva Biotecnologica*, **19**, 17 (2007).
3.   C.E. Barnes, *Industrial and Engineering Chemistry*, **30**, 449 (1938).
4.   H.S. Cockeram, L. S. *Journal of Cosmetic Science*, 12, 316 (1961).
5.   S. Fleix, S. Marianne, W. Tor, W.M. Bernd, *Pharmaceutical Technology*, **23**, 146 (1999).
6.   J.P. Ernest, *SHELLAC, Its production, manufacture, chemistry, analysis, commerce and uses*; Sir Isaac Pitman & Sons, Ltd.: London, 1935.
7.   S. Stummer, S. Salar-Behzadi, F.M. Unger, S. Oelzant, M. Penning, H. Viernstein, *Food Research International*, **43**, 1312 (2010).
8.   S. Limmatvapirat, C. Limmatvapirat, M. Luangtana-Anan, J. Nunthanid, T. Oguchi, Y. Tozuka, K. Yamamoto, S. Puttipipatkhachorn, *International Journal of Pharmaceutics*, **278**, 41 (2004).
9.   A.L. Campbell, B.L. Holt, S.D. Stoyanov, V.N. Paunov, *Journal of Materials Chemistry*, **18**, 4074 (2008).
10   A.L. Campbell,S.D. Stoyanov, V.N. Paunov, *Soft Matter,* **5**, 1019 (2009).
11.   S. Leick, M. Kott, P. Degen, S. Henning, T. Pasler, D. Suter, H. Rehage, *Physical Chemistry Chemical Physics*, **13**, 2765 (2011).
12.   J. Xue, Z.B. Zhang, *Journal of Microencapsulation*, 25, 523 (2008).
13.   J. Xue, Z.B. Zhang, *Journal of Applied Polymer Science*, **113**, 1619 (2009).

Mater. Res. Soc. Symp. Proc. Vol. 1498 © 2013 Materials Research Society
DOI: 10.1557/opl.2013.334

Biomimetic method for metallic nanostructured mesoscopic models fabrication

Gennady V. Strukov and Galina K. Strukova

Institute of Solid State Physics, Russian Academy of Sciences, 142432 Chernogolovka,
*E-mail: strukov@issp.ac.ru

ABSTRACT

Various metallic structures of complex shape, resembling natural objects such as plants, mushrooms, and seashells, were produced when growing nanowires by means of pulsed current electroplating in porous membranes. These structures occur as the result of nanowires self-assembling (biomimetics) if the electroplating is continued after the nanowires reach the membrane surface. By varying the membrane geometry and the pulsed current parameters, and alternating electroplating from two baths with different electrolytes, various models were fabricated, including a hollow container with wall thickness of 10-30 nm. The possibility of shape regulation for models was demonstrated: in certain conditions, mushroom- and shell-like convex-concave models of the same kind were obtained. The hierarchical structure of models at the nano-, micro- and mesoscopic levels is shown through fragmentation and chemical etching. This biomimetic method suggests an analogy between the shape-forming processes of natural plants and their metallic models. Nanostructured mesoscopic objects of metals (Ag, Pd, Rh, Ni, Bi), alloys (PdNi, PdCo, PbIn) as well as their combinations (PdNi/ Pb, PdNi/ PbIn) were obtained. The technological simplicity of the present method makes it suitable for fabricating nanostructured materials that may be efficient in catalysis, superhydrophobic applications, medical filters, and nanoplasmonics.

INTRODUCTION

In the process of nanowire growth by electroplating of metal on porous membranes, our attention was attracted by the mesoscopic structures forming on the top of the membrane when the growth was continued after the nanowires reached the membrane surface. The similarity between the obtained structures and natural objects (plants, mushrooms, shells) was striking. Development of bio-inspired methods of fabricating replicas or models of natural objects is a substantial research topic, as some natural materials possess unique properties still unattainable in artificial materials [1]. Superhydrophobicity of the lotus leaf and some other plants, and fracture- toughness of nacre [1,2], are among those useful properties. Recently, researchers' interest has been attracted by "nanoflowers", anisotropic metal nanostructures. For instance, the effects of giant Raman scattering on gold [3,4] and of catalytic activity on platinum nanoflowers [5,6] have been reported. However, comparison of our results and the data available in the literature provides no information on general methods that enable synthesis and shape control of such "plant" structures.

The present work aims is to demonstrate the method of controlled-growth, metallic nanostructured mesoscopic models as well as reveal their architecture and hierarchic structure. A particular issue is to show the feasibility of model shape control, i.e., fabrication of models with a certain shape type. Additionally, we will discuss possible applications of the method for creating novel materials with desired properties.

EXPERIMENT

We prepared metallic mesostructures on porous membranes via pulsed current electroplating using a simple two-electrode scheme. One surface of the membrane was covered with 50-100 nm of copper layer that served as a cathode in the process of electrolysis. In order to grow nanowires only through the pores, the cathode backside was insulated to prevent any undesired electrical contact with the electrolyte. Platinum foil placed several millimeters away in front of the membrane was used as an anode. Cathode and anode, electrically linked to the pulsed current generator, were placed in a bath with electrolyte to undergo a cycle of rectangular current pulses. The number of current pulses, their amplitude and length, as well as the duration of pauses between pulses, were all computer controlled. The pulse frequency was in the range of 30-166 Hz, while the duty cycle was varied between 50 and 99 %. The automatic setup for alternating (layer by layer) electroplating from two baths [7] allowed us to obtain volume hybrid mesostructures. Electrolyte composition and other details of the experiment are presented in our papers [7,8]. The architecture and hierarchic structure of the "shells" were studied using images taken both before and after mechanical fragmentation either in an ultrasonic bath or by chemical dissolution (etching). SEM tests on samples were carried out with SUPRA 50VP and JEOL scanning electron microscope.

RESULTS

Various "vegetable" nanostructured mesoscopic models were formed at the anodic aluminum oxide (AAO) membrane with a disordered arrangement of pores due to self-assembly of metallic nanoclusters and nanowires, which occurs if the electroplating is continued after the nanowires have reached the membrane surface. Figure 1 shows some of the PdNi alloy mesostructures, including classical Mandelbrot fractals-"cauliflower", "broccoli", "algae".

Figure 1. Mesoscopic models of PdNi alloy: a) "cauliflower" and "algae", b) "vegetable" structures at different stages of growth, c) "brooms" and "broccoli", d) "cactus", e) "coniferous branches", f) and g) "bouquets", h) "patty-pan squash".

Graceful plants made of Ag are presented in Figure 2. The structure of Ag "branches and berries" (2a) contains spherical or elongated nanostructured granules of silver sized 20 to 40 nm ("berries"). On the other hand, the structure of the Ag "branch with leaves" (2c) is decorated with swept flat "leaves".

Figure 2. Mesoscopic models of Ag: a) "branch with berries", b) distinctive "leaf", c) "branch with leaves".

Metallic hybrid mesostructure PdNi/Pb was grown by means of a special technique of alternating electroplating in two baths with corresponding electrolytes. Figure 3 presents one of those models and its fragments: long straight "palm branches" (3a) with straight leaves"(Figure 3b), each tipped with a wonderful sedum-like bud, which, in its turn, is a hollow pipe with wall thickness of 10-30 nm (Figure 3c).

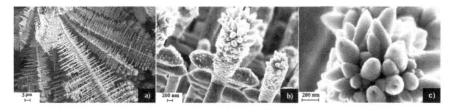

Figure 3. "Flower wreath" formed by alternate electroplating of Pb and PdNi alloys: a) general view of the structure, b) a single "leaf", c) a single "bud".

A similar procedure was used to grow a volume hybrid mesostructure PdNi/PbIn "fern". Figure 4 shows this model and its fragments. Arrow-like leaves formed by four orthogonal blades are situated along the stem; the blades consist of alternating submicron columns which are formed by 100-200 nm spherical grains.

Figure 4. "Fern" formed by alternate electroplating of PbIn and PdNi alloys. a) general view, b) fern "leaves", c) "fern stem".

The combination of specifically treated polymer membranes and the pulsed current regimes allowed reproducible growth of convex-concave structures of a certain type, such as "lotus leaves" or "seashells". Figure 5 shows "seashells" in rows that are reproducibly obtained under particular growing conditions. The convex-concave PdNi models alter their appearance due to slight varying of the pulsed current regime (Figure 6 a-d, f). Under similar conditions an Ag model resembling a cabbage leaf was grown (Figure 6e).

Figure 5. Shells in rows.   **Figure 6.** Typical convex-concave Pd/Ni and Ag (e) models.

Similar structures were obtained from copper, nickel, palladium, rhodium, and PdCo alloy.

The obtained structures have several characteristic features. The growth of the shell walls starts from its "root". The "root" ("bottom") of the shell is a round or oval area generated by the beams of layered wires and can be compared to a stump. The edge of this area is a circle, an oval or an unclosed horseshoe-like line. Bundles of nanowires growing from separate points on this line form the walls of the shells. Structurally, they resemble bunches branching upward. Such "brooms", growing from separate points, were revealed among various volume PdNi alloy structures as shown in Figure.1.

The ends of the nanowire bundles often rise to the outer "shell" surface as "nanoflowers" that, in turn, can serve as templates for growing new nanowires. The inner surface of the "shell" or "lotus leaf" has a characteristic micro-surface pattern with a nano-level roughness. This nonuniform surface pattern is an entwined-like structure with its lines directed from the "root" to the periphery and weakly pronounced transverse terraces (Figure 7a).

This pattern can be considered as a manifestation of the inner "shell" architecture. Such architecture is observed if the surface layer has been removed mechanically in an ultrasonic bath or by means of chemical etching (Figure 7,b,c).

Figure 7. Inner Pd-Ni "shell" surface: (b) after 10 s, and (c) after 30 s of chemical etching.

The metallic "shell" has a bowl-shaped frame; its walls are composed of layers of densely packed nanoelements. Each nanoelement is a conical bundle of nanowires grown as a "wine glass". The shape of the "shell" frame (Figure 6) copies the shape of the fragments and of the nano- building elements, that is, the conical nanowire bundles (Figures 7,8). I.e., the hierarchic structure is reproduced at the nano-, micro- and mesoscopic levels.

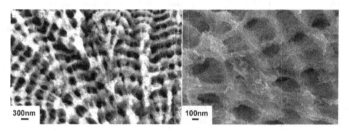

Figure 8. Etching pattern of the inner "shell" surface: fractal networks with circular bands ("bandages").

DISCUSSION

Numerous investigations have shown that the excellent mechanical properties of natural materials are due to their complex architecture and hierarchic structure at the nano-, micro- and mesoscopic levels [2]. The natural nacre hierarchical scheme is called "bricks and building mortar". The superb strength, viscosity, and resistance of this material are achieved by the reliable nanostructure of alternating layers of natural organic polymer, protein (10 -50 nm), and aragonite pellets (200-900 nm) 5-8 μm in diameter. In our case, the grown convex-concave models also have hierarchic structure with the resulting architecture based on three-dimensional lattice. Obviously, a detailed study of these nano-structures by means of more powerful electron microscopy is required. The first TEM data revealed the essential distinction of the crystal state and phase composition of the "bricks" and the layer between them. This suggest that the use of various modes of pulse current and electroplating of various metals from two baths enables the implementation of a biomimetic approach in creating a model of natural nacre. We have managed to grow "shells" from Pd-Co alloy and copper and that extends the set of feasible materials. Yet, the question is still open: What model sizes can be grown by the proposed method so that the hierarchic structure is preserved at the nano-, micro- and mesoscopic levels? So far we have grown a 4×2 mm "shell" from Pd-Ni alloy.

CONCLUSIONS

We have proposed a method of fabricating metallic mesoscopic models of natural objects-plants, mushrooms, shells- by programmable pulsed current electroplating on templates. Variation of the template form and the pulsed current parameters, and the use of different electrolytes, made it possible to grow an impressive diversity of structures, including hollow structures with 10-30 nm wall thickness. The feasibility of model shape control is shown: the conditions for growing definite convex-concave, lotus leaf-, mushroom-, and shell-like structures have been found. The internal hierarchic fractal structure of the grown models composed of nanowire bundles has been revealed. Thus, the method proposed ensures not only external similarity of the models to natural objects but also their inherent hierarchic structure at the nano-, micro- and mesoscopic levels, i.e., it is a biomimetic method. On the other hand, it also suggests that pulse growth on templates is a tool of morphogenesis of numerous mushrooms and plants. Nanostructured models have been grown from normal (Ag, Pd, Rh), magnetic (Ni, Pd-Ni, Pd-Co) and superconducting (Bi, Pb-In, Pb-Bi) metals and from alloys, which opens up prospects for their use in the creation of nanodevices. Nanostructured large surface models such as "silver wood", and similar structures from other metals, are interesting for applications such as catalytic filters, batteries, and supercapacitors. The method proposed enables the growth of nanostructured composites of metal-semiconductor (cadmium sulfide) mesostructures. Such a material holds promise for a practical use in solar energy production via plasmon-enhanced catalysis [9]. An attractive prospect is self-assembly growth of target hybrid structures for high-performance electronic devices. Nanowire brush-like coatings may be useful as materials for microengineering. Hollow metallic structures with an opening are good candidates for applications as nano-containers and nano-reactors.

ACKNOWLEDGEMENTS

The authors thank colleagues from the Laboratory for Superconductivity Sergey Egorov and Ivan Veshchunov for their help in this work.

REFERENCES

1. Xianfeng Wang, Bin Ding, Jianyong Yu, Moran Wang. Nano Today, Vol. 6, Issue 5, 510-530, 2011.
2. Corni, T.J. Harvey, J.A. Wharton, K.R. Stokes, F.C. Walsh and R.J.K. Wood. Bioinspiration & biomimetics, 7 , pp 031001-24 (2012).
3. Dan Xu, Jiangjiang Gu, Weina Wang, Xuehai Yu, Kai Xi and Xudong Jia. Nanotechnology, 21, Issue 37, p.5101 (2010).
4. Linmei Li and Jian Weng. Nanotechnology, 21 , Issue 30, p. 5603 (2010).
5. Xiaomei Chen , Bingyuan Su , Genghuang Wu , Chaoyong James Yang , Zhixia Zhuang , Xiaoru Wang and Xi Chen. J. Mater. Chem., 22, 11284-11289 (2012).
6. Hongmei Zhang, Weiqiang Zhou, Yukou Du, Ping Yang and Chuanyi Wang. Solid State Sciences, Volume 12, Issue 8, 1364-1367, 2010.
7. G.V. Strukov, G.K. Strukova, E.D. Shoo, S.I. Bozhko, Yu.P. Kabanov. ISSN 0020-4412, Instruments and Experimental Techniques 52 (5) (2009) 727–730; Original Russian Text published in Pribory i Tekhnika Eksperimenta 5 (2009) 123–126.
8. G.V. Strukov, V.S. Stolyarov, G.K. Strukova, V.N. Zverev. Physica C, 483,162-164, 2012.
9. Suljo Linic, Phillip Christopher and David B. Ingram. Nature Materials, 10, 911–921 (2011).

Mater. Res. Soc. Symp. Proc. Vol. 1498 © 2013 Materials Research Society
DOI: 10.1557/opl.2013.335

Microfluidics via Controlled Imbibition

Ville Jokinen[1] and Sami Franssila[1]
[1]Aalto University, School of Chemical Technology, Department of Materials Science and
Engineering, Tietotie 3, 02150 Espoo, Finland.

ABSTRACT

Strategies for controlled imbibition utilizing both topographical cues as well as patterned
surface chemistries are presented. Triangle-like microstructures were used for directional wetting
of liquids into a limited sector of the surface. Chemical patterns on top of a nanopillar geometry
allowed the patterning of both water and oil droplets. Applications of controlled imbibition in
dried droplet mass spectrometry as well as liquid-liquid extraction are also presented.

INTRODUCTION

Controlled imbibition is a way of realizing capillarity driven microfluidics on open
surfaces. Two basic approaches exist for controlling the imbibition: surface topography (micro
and nanostructures) and patterned surface chemistry. There has been rapid progress in utilizing
both of these approaches during the last 5 years. Surfaces with a uniform surface chemistry but
specially tailored surface topography have been used to achieve e.g. polygonal imbibition [1,2]
and directional wetting [3-6]. On the other hand, surfaces with patterned surface chemistry have
been used for controlling the shape of liquid droplets (and hence, the direction of imbibition)
[7,8] as well as for achieving directional wetting properties [9].

EXPERIMENT

Surfaces for directional imbibition [3] were made out of SU-8 photoresist (SU-8 50,
Microresist technology). The contact angles of the SU-8 surfaces were modified with oxygen
plasma (Plasmalab 80+, Oxford Instruments) to 15°-50°. Surfaces with chemical patterns
[7,8,10,11] were fabricated out of silicon nanograss and a patterned fluoropolymer coating.
Silicon nanograss was created by cryogenic deep reactive ion etching (Plasmalab System 100,
Oxford Instruments. The surface was then coated with a CHF_3 plasma fluoropolymer. Photoresist
masking was used to selectively modify the surface into hydrophilic by etching the
fluoropolymer and oxidizing the underlying silicon nanograss with oxygen plasma. Contact
angles were measured using the sessile droplet method (Cam 101, Biolin Scientific) with droplet
sizes 1 µl - 10 µl. Sliding angles were measured with an in house built apparatus from 10 µl
droplets. The imbibition experiments reported were performed in open atmosphere with 45% ±
5% air humidity. The exception was the evaporation experiments performed for mass
spectrometry application, where the droplets evaporated in normal laboratory ambient with
unknown air humidity. Imaging mass spectrometry (Autoflex III, Bruker Daltonics) of peptide
samples was performed with a 25 µm grid and ≈ 50 µm laser spot size. The test analyte was des-
arg[9]-bradykinin (Peptanova). Liquid-liquid extraction was studied with fluorescence microscopy
(Zeiss AxioScope A1, Carl Zeiss Oy). The test analyte was fluorescein and donor, acceptor and

organic phases were 98mM HCl, 40mM NaOH and Octanol, respectively. All chemicals for the liquid-liquid extraction experiment were from Sigma-Aldrich.

RESULTS AND DISCUSSION
Directional wetting via asymmetric surface microtexture

Directional wetting based on triangle-like SU-8 microtexture [3] is demonstrated in Figure 1. The structure presented has a depth of 15 μm, advancing contact angle of 40° on reference planar surface and the triangle dimensions are 20 μm x 80 μm. We observed that the distance that the liquid meniscus reaches out from a row of triangular microstructures is dependent on whether the liquid is leaning on the bases or the tips of the triangles. Fig 1a and Fig 1b show the measurement of this effect. An auxiliary row of micropillars was utilized to measure the reach with accuracy of ± 1 μm (2 μm discretization) by observing whether the liquid meniscus makes contact with the auxiliary pillars. Fig 1a, left side, demonstrates that the meniscus leaning on the tips of the triangles makes contact with the auxiliary pillar at 9 μm distance but not with the pillar at 11 μm distance, so the reach is 10 μm ±1 μm. Similarly, the reach of the meniscus leaning on the bases of the pillars is shown to be 20 μm ±1 μm in Fig 1b, right side. This effect can then be translated into directional wetting by making the distance between pillar rows to be some value between the reach from the tips and the reach from the bases. Fig 1c (entrance of the channel) and Fig 1d (end of the channel) demonstrate this effect with a 14 μm spacing; in the unfilled direction the liquid does not advance from the first row while in the filled direction the whole test structure (100 rows, 1 cm) is filled. In effect, this means that the liquid only spreads to a 180° sector. We have also demonstrated that the effect can be used to achieve spreading in a 90° sector (Fig 1e and Fig 1f). The shape of these structures is shown in the inset of Fig 1e.

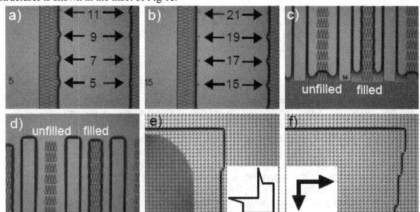

Figure 1. Directional wetting via asymmetric, triangle-like microstructures. a) Reach of liquid meniscus from the tips micropillars. b) Reach of a liquid meniscus from the bases of micropillars. c), d) Unidirectional imbibition utilizing triangular micropillars (imbibition in 180° sector). e), f) Unidirectional imbibition into a 90° degree sector. The insets show the shape of this micropillars, and the allowed imbibition directions (toward right and bottom).

Controlled imbibition via surface chemistry patterns

Surface chemistry patterns on top of silicon nanograss topography can be utilized for controlled imbibition of water [7] and also oils [8]. Figures 2a and 2b show patterned water droplets on chemically modified silicon nanograss. The hydrophilic areas that the droplets are sitting on have an advancing contact angle of 0°, while the hydrophobic areas surrounding the droplets have both advancing and receding water contact angles near 170°. Two parameters are important for creating accurately patterned water droplets, the contact angle contrast between the hydrophilic and hydrophobic areas (≈170° in our case) and the contact angle hysteresis of the hydrophobic area (<5° in our case). The contact angle contrast is the parameter that allows controlled imbibition and it also determines how accurately certain problematic features, such as sharp corners and concave areas with small radii of curvature, can be patterned. The low hysteresis of the hydrophobic side helps during the application of the droplet as it ensures that the droplet can completely recede away from the initial site.

Patterning oil droplets is more challenging compared to water due to the unavailability of surface chemistries that are inherently oleophobic. Oil repellent surfaces are fabricated through re-entrant geometries, but these would be very difficult to fabricate in the lateral direction as required by droplet patterning. We have been able to pin the advancing oil meniscus at selected locations with surface chemistry patterns on top of micropillar and nanopillar geometries [8] (Figures 2 c-e), but the approach has been limited to oils with surface tensions higher than ≈ 30 mJ/m^2, corresponding to nominal contact angles ≈ 55° or higher between the oil and the fluoropolymer.

Figure 2. Controlled imbibition via surface chemistry patterns, a) Triangular water (colored) droplets with sharp tips. b) Torus shaped water droplets. c), d), e) Optical micrograph of various test geometries on a patterned oleophobic/oleophilic surface. The oil used was olive oil with a contact angle 57° with planar reference surface with fluoropolymer chemistry.

Applications of controlled imbibition

Applications for controlled imbibition include e.g. liquid transport and controlled deposition. Figure 3 presents three applications that we have demonstrated. Figures 3a and 3b demonstrate a unidirectional droplet track. The structure consists of a long unidirectional channel and multiple liquid application sites. The structure could be used e.g. for selective surface treatments of various parts of the main track.

Figures 3c and 3d present a dried droplet solute deposition application [10]. In surface assisted laser desorption ionization mass spectrometry [12], the analyte is first deposited on the surface from a drying liquid droplet. We noticed that we can use a flower and petals shaped geometry to split and concentrate the sample into the peripheral spots, while practically no analyte remained in the center of the droplet after drying.

Figures 3e-3h present a liquid-liquid-liquid extraction application based on designer multiphase droplets [11]. A multiphase droplet consists of two or more connected immisicible liquid phases. Figure 3f shows a three phase, aqueous-organic-aqueous, droplet. Such geometries can be used for liquid phase extraction of analytes from one aqueous phase to another through the organic phase. These surfaces could be used as rapid, low volume extraction platforms for analytical chemistry, and could also be utilized as a pretreatment strategy for other miniaturized analysis devices.

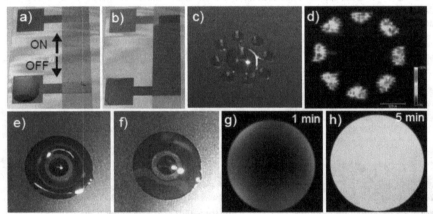

Figure 3. Applications of controlled imbibition. a) A unidirectional liquid track with multiple liquid application sites on the side (2 shown). Liquid is applied on the bottom application site. The direction of imbibition is marked. b) The imbibition reaches the next liquid application site c) A water droplet with a flower petal geometry for dried droplet applications. d) The resulting dried droplet solution pattern as seen by mass spectrometric imaging. The analyte solution in this case was 1μM des-arg[9]-bradykinin. e), f) A multiphase droplet before (e) and after (f) application of the organic phase. The droplet consists of two colored water phases separated by a clear hexadecane organic phase. g) h) Multiphase droplet as a liquid-liquid-liquid extraction platform. The analyte is extracted from the outer aqueous phase to the central aqueous phase through the organic phase. The analyte is 40μM fluorescein.

CONCLUSIONS

Surface topography, surface chemistry and combinations can all be used for various types of controlled imbibition. Due to the range of possible geometries, dimensions, chemistries, liquids and applications, the field has plenty of room to grow.

ACKNOWLEDGMENTS

The authors thank Ms Marianne Leinikka for help with the directional wetting experiments and Dr Lauri Sainiemi for assistance with the surface chemistry patterned silicon nanograss surfaces.

REFERENCES

1. C. W. Extrand, S. I. Moon, P. Hall and D. Schmdidt, *Langmuir* **23**, 8882-8890, (2007).
2. L. Courbin, E. Denieul, E. Dressaire, M. Roper, A. Ajdari, H. A. Stone, *Nature Mater.* **6**, 661-664, (2007)
3. V. Jokinen, M. Leinikka and S. Franssila, *Adv. Mater.* **21**, 4835-4838 (2009).
4. K-H. Chu, R. Xiao and E. N. Wang, Nature Mater, **9**, 413-417 (2010).
5. T. Kim amd K. Y. Suh, *Soft Matter* **5**, 4131-4135, (2009).
6. N. A. Malvadkar, M. J. Hancock, K. Sekeroglu, W. J. Dressick and M. C. Demirel, *Nature Mater.* **9**, 1023-1028, (2010).
7. V. Jokinen, L. Sainiemi and S. Franssila, *Adv. Mater.* **20**, 3453-3456 (2008).
8. V. Jokinen, L. Sainiemi and S. Franssila, *Langmuir* **27**, 7314-7320 (2011).
9. C. Q. Lai, C. V. Thompson and W. K. Choi, *Langmuir* **28**, 11048-11055 (2012).
10. V. Jokinen, S. Franssila and M. Baumann, *Microfluids. Nanofluids.* **11**, 145-156 (2011).
11. V. Jokinen, R. Kostiainen and T. Sikanen, *Adv. Mater.* in press doi:10.1002/adma.201202715, (2012).
12. V. Jokinen, S. Aura, L. Luosujärvi, L. Sainiemi, T. Kotiaho, S. Franssila and M. Baumann, *J. Am. Soc. Mass Spectrom.* **20** 1723-1730 (2009).

Functional and Responsive Materials Exploiting Peptide and Protein Self-Assembly

Mater. Res. Soc. Symp. Proc. Vol. 1498 © 2013 Materials Research Society
DOI: 10.1557/opl.2013.336

Spider silk morphology for responsive materials

Juan Guan, David Porter*, and Fritz Vollrath

Department of Zoology, South Parks Road, University of Oxford, Oxford, OX1 3PS, UK

*Correspondence: david.porter@zoo.ox.ac.uk

ABSTRACT

This study reveals that an "old" mechanism for shape memory in oriented polymers is in fact just one separate contribution for "supercontraction" in *Nephila* spider major ampulate silks. When *Nephila* spider silks are in contact with liquid water, they "super"-contract up to 28% of the original stretched length. However, we discovered that under glass transition conditions these silks only relax with a maximum shrinkage of 13%, and this phenomenon is defined as T_g-contraction. Structural components permanent order (PO), permanent disorder (PD), meta order (MO) and meta disorder (MD) were proposed from the primary amino-acid sequence of the silk protein to explain morphological changes in the two contraction phenomena: MD contributes 13% of the full supercontraction and contributes to T_g-contraction; whereas MO (the proline-containing motifs) contributes the rest for the full super-contraction and does not contribute to T_g-contraction. The morphology in *Nephila* spider silk structure suggests two separate mechanisms to generate the shape memory effect in synthetic polymers.

INTRODUCTION

Spider silk is a natural protein fibre with outstanding mechanical properties, and its high strength combined with large extensibility outshines many synthetic materials. Recently, the thermal conductivity [1], shape memory [2], and humidity-driven changes in mechanical performance [3] have raised new interest in this promising future material.

Another remarkable property of spider dragline silk is "super-contraction" [4]: when dragline silk is in contact with liquid water, it contracts instantaneously and often substantially depending on the species of the spider and silk composition and morphology. More interestingly, when the contracted silk is restretched back to its original length, it retains its ability to super-contract. The contraction-stretching cycles can be reproduced many times [3]. Supercontraction was first discussed and defined by Robert Work [5] and now this phenomenon provides new potential for spider silk applications; for example, as tunable shape memory polymers, which are of great technological interest [6].

On supercontraction, the phenomenon itself, there have been substantial observations. However, in respect to the morphology of spider silk and structural contributions to supercontraction, there is not yet a complete picture. Based on a thorough literature study, we proposed two structural relaxations towards supercontraction correlated with two morphologies: (i) the oriented disordered domains and (ii) the proline-rich ordered domains. The oriented disordered domains, so-called meta-disorder (MD) are able to relax under glass transition conditions. This partial contraction is compared with full supercontraction, so-called T_g-contraction, in this paper.

EXPERIMENT

A medium size *Nephila edulis* spider was firstly anesthetized by CO_2, and then immobilized by pinning onto a platform. Dragline threads were then identified under a microscope. Once the spider recovered, draglines were forcibly reeled at a controlled speed of 10 mm/s and collected on our standard rotating spool with horizontal spaces and automatic advance. The silks directly from the spool are categorised as virgin silks. They were carefully transferred from the spool to paper holders. Supercontracted silks were obtained by immersing the virgin silks which were transferred first from spool onto dividers. The maximum shrinkage was recorded for each silk. Both natural and supercontracted silks were kept restrained on paper holders in the testing room (temperature 20 °C and relative humidity < 40%) before being tested.

The diameters of virgin silks were measured on SEM (Jeol Neoscope JCM-5000). The average value from the measurements was used for analyzing the results of mechanical testing. Tension was carefully kept when virgin silks were transferred from the spool to the SEM stub. For supercontracted silks, we assumed a constant volume rule between virgin silks and supercontracted silks [7], hence the diameter was estimated to be 7.7 microns when contracted by 30% along the fibre axis, and this was validated by a few direct measurements. For the partially contracted fibres, the constant volume rule was again applied to estimate the cross-sectional areas.

DMTA Q800 (Dynamic Mechanical Thermal Analysis, TA Instrument) is used for both quasi-static and dynamic mechanical testing. A humidity accessory was adapted to control the humidity-temperature experiments. The resolution of temperature was 1 °C and relative humidity was 2%. Static tensile tests were conducted at a force-ramp rate of 0.05 N/min (comparable to a strain rate of 1% s^{-1}, at room temperature after being equilibrated in the chamber for 10 minutes under nitrogen purge. Dynamic mechanical test on supercontracted silks (single fibre) was set up as: 50 MPa static stress, ~0.2% dynamic strain, 1 Hz frequency, 50 °C, relative humidity ramp from 0% to 90% at 1%/min rate. Length tests were conducted in controlled-force mode. For the temperature set, virgin silks were kept under static stress of 10 MPa through humidity of 0% to 90% at five temperatures 10, 25, 37.5, 50 and 70 °C. For the stress set, different static stresses from 0 MPa to 20 MPa were applied to virgin silks through changing humidity from 0% to 90%. Length and force were live monitored in all the tests.

RESULTS AND DISCUSSION

Glass transition

Figure 1 shows the storage modulus and loss tangent of supercontracted dragline silk change as a function of ramping humidity at a temperature of 50 °C. As the humidity increases to 63%, the storage modulus drops more rapidly down to below 1 GPa and at the same time the loss tangent shows a peak at RH 82%. The combination of temperature and humidity at 50 °C and 63% agrees with the glass transition conditions observed by Fu and Plaza for *A.pernyi* silkworm silk and *Argiope* spider silks [8,9]. The graph shows the amorphous structure in spider silk becomes mobile at the glass transition.

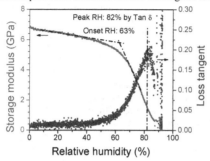

Figure 1. Dynamic mechanical properties of supercontracted Nephila spider silk through ramping humidity at 50 °C.

Linking glass transition to contraction

Because virgin silk contracts or supercontracts within the humidity and temperature regime (which interfere with the pure glass transition effect), we cannot directly measure the glass transition in virgin silks in the same experiment as that for supercontracted silk in Figure 1. However, by using fully-supercontracted silk, we know the disordered structure is able to relax at the glass transition, and by assuming the disordered structure does not change through supercontraction we confirm that the same combination of temperature and humidity should induce the glass transition in virgin *Nephila* dragline silks. The direct comparison of modulus change in supercontracted silk and length change in the virgin silk in Figure 2 shows the same transition, and hence a clear link between the two events.

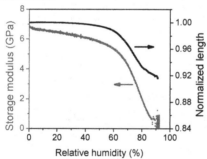

Figure 2. Glass transition in supercontracted silk and length contraction in virgin silk through ramping humidity.

T_g-contraction

Figure 3(a) shows how varied temperatures affect the contraction of virgin dragline silk at a constant static stress of 10 MPa. It shows the glass transition happens at lower humidity for higher temperature. However, independent of the temperature, the maximum shrinkage seemed to be able to reach a consistent plateau of about 9% at this stress if all runs were allowed to reach high humidity.

Since the shrinkage observed in Figure 3(a) is far lower than the full supercontraction shrinkage of 30%, the effect of restraining stress on T_g-contraction was investigated next. Figure 3(b) shows that stress does affect the contraction value, but not the onset humidity. With lower restraining stress, the silk is able to contract more. When the stress goes down to 1 MPa, the shrinkage seems to reach a maximum value of 13%, which is still lower than the full potential supercontraction.

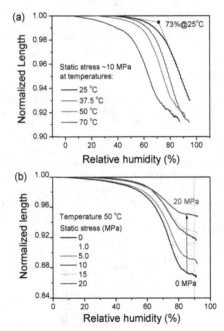

Figure 3. Length contractions of *Nephila* spider silk: temperature effect (a) and stress effect (b).

Morphology and structural contributions

In order to correlate the structure with supercontraction and T_g-contraction for *Nephila* dragline silks, it is necessary to first quantify the main structural and morphological components. Here we discuss a simplified structural model, in which the different domains act independently of each other. We define the structural components in spider silk as permanent order (PO), meta-order (MO), meta-disorder (MD) and permanent disorder (PD), loosely correlating to beta-sheet, proline-containing beta or helical structure, oriented random coil or helical structure, and random coil. From the protein sequence (*Nephila clavipes* Spideroin 1 and 2), we take the 30% alanine segments as PO and 20% of the end groups as PD as suggested by references [10, 11]. Porter and Vollrath also suggested that the 21% –GPGXX– segments (derived from 4.2% proline, and X represents any amino acid other than glycine or proline) can be effectively oriented and ordered under load as the torsional twists of the proline segments are pulled into line to allow the other 4 peptide segments to order and be held in place by strong hydrogen bonding [10]. The rest of 29% –GXG– segments are conventionally associated with disordered amorphous chains, however, they can be oriented under load and can relax under glass transition conditions, hence are categorized as MD.

In terms of morphology, PO or PD is not susceptible to structural shrinkage, while MO and MD can both be contributors to contraction, but each operating with a different mechanism. MO component contracts from beta sheet to condensed sphere due to the twisting of the proline segments, with a maximum shrinkage of about 86% for the -GPGXX- blocks (estimated from the dimensions of the 30 to 35 peptide segment blocks). We further suggest that MD fraction contracts to alpha-helix or 3_{10}-helix form with a maximum shrinkage of about 50%.

Now we can make the link between structural components and contraction events. Table 1 summarizes the proposed protein sequence, structural components, and the morphological contractions. In *Nephila* dragline silk, the 29% MD component –GXG– can contract up to 15% and the 21% MO component –GPGXX– can contract up to 18%, for a total of 33%, which is in good agreement with published values of 0.28 to 0.33 for the total, and our experimental values of 13% for T_g-contraction and 28% for the total supercontraction.

Table 1 Fractions and contributions to the supercontraction of the different structural and morphology components in *Nephila edulis* fibres.

Component	Fraction	Structure	Maximum contraction	*Nephila edulis*
Permanent order, PO	0.30	-A$_n$-	0	0
Permanent disorder, PD	0.20	Long side chains?	0	0
Meta order, MO	0.21	-GPGXX-	0.86	0.18
Meta disorder, MD	0.29	-GXG-	0.50	0.15

CONCLUSIONS

The two mechanisms for "super-contraction" in spider silk were attributed to MD oriented random coil for T_g-Contraction, and MO stretched helical for the rest of the full contraction. This offers two routes to design new responsive materials to environmental stimuli, such as humidity, temperature and mechanical stress. We also demonstrate that the new approach of using dynamic mechanical thermal analysis, DMTA, to understand the structure-property relations of micron-thick natural silks is a powerful analytical tool that is potentially useful for other natural materials, particularly with temperature-humidity scans to augment the standard temperature scans.

ACKNOWLEDGMENTS

Thanks to the European Research Council (MASP2-GA-2008-233409), AFOSR (F49620-03-1-0111) and Chinese Ministry of Education-University of Oxford Scholarship (for JG) for financial support, and Dr. Chris Holland for valuable advice as well as DMA training for Juan.

REFERENCES

1. Huang, X.; Liu, G.; Wang, X., New Secrets of Spider Silk: Exceptionally High Thermal Conductivity and Its Abnormal Change under Stretching. *Adv. Mater.* **2012,** *24* (11), 1482-1486.
2. Emile, O.; Le Floch, A.; Vollrath, F., Biopolymers: Shape memory in spider draglines. *Nature* **2006,** *440* (7084), 621-621.
3. Agnarsson, I.; Dhinojwala, A.; Sahni, V.; Blackledge, T. A., Spider silk as a novel high performance biomimetic muscle driven by humidity. *J. Exp. Biol.* **2009,** *212* (13), 1989-1993.
4. Guan, J.; Vollrath, F.; Porter, D., Two Mechanisms for Supercontraction in Nephila Spider Dragline Silk. *Biomacromolecules* **2011,** *12* (11), 4030-4035.
5. Work, R. W., A comparative study of the supercontraction of major ampullate silk fibres of orb web building spiders (Araneae). *J. Arachnol.* **1981,** *9*, 299-308.
6. Behl, M.; Lendlein, A., Shape-memory polymers. *Materials Today* **2007,** *10* (4), 20-28.
7. Guinea, G. V.; Perez-Rigueiro, J.; Plaza, G. R.; Elices, M., Volume Constancy during Stretching of Spider Silk. *Biomacromolecules* **2006,** *7* (7), 2173-2177.
8. Fu, C. J.; Porter, D.; Shao, Z. Z., Moisture Effects on Antheraea pernyi Silk's Mechanical Property. *Macromolecules* **2009,** *42* (20), 7877-7880.
9. Plaza, G. R.; Guinea, G. V.; Perez-Rigueiro, J.; Elices, M., Thermo-hygro-mechanical behavior of spider dragline silk: Glassy and rubbery states. *J. Polym. Sci. Pol. Phys.* **2006,** *44* (6), 994-999.
10. Vollrath, F.; Porter, D., Spider silk as an archetypal protein elastomer. *Soft Matter* **2006,** *2*, 377-385.
11. Liu, Y.; Sponner, A.; Porter, D.; Vollrath, F., Proline and processing of spider silks. *Biomacromolecules* **2008,** *9* (1), 116-121.

Mater. Res. Soc. Symp. Proc. Vol. 1498 © 2013 Materials Research Society
DOI: 10.1557/opl.2013.337

Morphology Control of Alzheimer Amyloid β Peptide (1-42) on the Multivalent Sulfonated Sugar Interface

Yoshiko Miura[1] and Tomohiro Fukuda[2]

[1]Department of Chemical Engineering, Kyushu University, 744 Motooka, Nishi-ku, Fukuoka, 819-0395, Japan
[2]Department of Applied Chemistry and Chemical Engineering, Toyama National Collage of Technology, 13 Hogo-machi, Toyama city, Toyama, 939-8630, Japan

ABSTRACT

The amyloidosis of amyloid β (1-42) was investigated by the well-defined glyco-cluster interface. We prepared monovalent, divalent, and trivalent 6-sulfo-N-acetyl-D-glucosamine immobilized substrates. The interaction between amyloid β and 6-sulfo-N-acetyl-D-glucosamine was amplified by multivalency of divalent and trivalent 6-sulfo-N-acetyl-D-glucosamine. The morphology of amyloid β were investigated by AFM, and we found the morphology of amyloid β aggregates were determined by the kinds of displayed saccharide-valency. Amyloid β had tendency to form spherical objects on the multivalent 6-sulfo-N-acetyl-D-glucosamine, but form fibrils on the monovalent 6-sulfo-N-acetyl-D-glucosamine. Spherical amyloid β was more toxic than fibrillar amyloid β to HeLa cells. These results suggested that the multivalency of was significant in its morphology and aggregation effects at the surface of the cell membrane mimic.

INTRODUCTION

Alzheimer's disease (AD) is a serious dementia, accompanied by the presence of senile plaques [1]. The main components of senile plaques are amyloid beta (Aβ) peptides of Aβ (1−40) and Aβ (1−42). Since aggregates of Aβs exist in AD patients, it is believed that Aβ aggregates injure the neuron cells. The onset of AD is related to amyloidosis of Aβ, and Aβ aggregates generally form insoluble fibrils [2, 3]. However, recent research indicated that the morphology of Aβ aggregates varied, and the cytotoxicity of Aβ has been reported to relate to the morphology and aggregation of Aβs as protofibrils, amyloid-beta derived diffusible ligand (ADDL), and amylospheroids [4-6]. We were interested in the reason for the change in morphology of Aβ and the relationship of the morphological change to the cytotoxicity of Aβ. It has been reported that amyloidosis is induced by the interaction of Aβ with the cell surface. Various sugars on cell surfaces have been reported to be involved in the amyloidosis of Aβ. It is well-known that interactions of Aβ with glycosaminoglycans such as heparin and ganglioside GM1 clusters are also closely related to amyloidosis [7, 8]. Meanwhile, the recognitions where sugars were involved were impossible to regulate due to the clustering effect. Consequently, it is possible that the morphologies and biofunctionalities of Aβ are affected by the degree of sugar clustering. However, detailed analyses of amyloidosis with well-defined glyco-clusters have not been reported.

In this study, we investigated the influence of a sugar on the aggregation behavior and cytotoxicity of Aβ with well-ordered glyco-clusters at the surface of a cell membrane mimic. As the target, we selected Aβ (1−42) that is a predominant biomolecule in the early pathological stage of senile plaque formation. We chose 6-sulfo-N-acetyl-D-glucosamine (6S-GlcNAc), which is the most abundant structure of heparin and has a similar structure to sialic acid, as a representative sugar for this study. We prepared functional self-assembled monolayers (SAMs) with glyco-clusters of mono (**1**)-, di (**2**)-, and tri (**3**)-valent 6S-GlcNAc (Figure 1) on a gold substrate via click chemistry [9]. We investigated the aggregation behavior, properties, and cytotoxicity of Aβ (1–42) on these SAMs.

EXPERIMENTAL

The multivalent sugar compounds of 6S-GlcNAc (**1-3**) were synthesized, and confirmed by ^{1}HNMR and mass spectra. The acetylene-terminated disulfide was synthesized following the previous method. The substrate was first modified by self-assembled monolayer (SAM) with acetylene-terminated disulfide. The multivalent 6S-GlcNAc was immobilized on the SAM by Click Chemistry with CuSO$_4$ and ascorbic acid.

The interaction between 6S-GlcNAc and Aβ (1-42, peptide institute) was measured by SPR of Biacore 3000 (GE Healthcare) with phosphate buffer solution (0.1 M NaCl, 20 mM potassium phosphate buffer (pH 7.4), 0.008% NH$_3$). The morphology of Aβ was measured by AFM (SPI3800N, Seiko Instruments). The cytotoxicity of Aβ was measured using HeLa cells, where the cell was cultivalted on the sugar displays of **1-3** with addition of Aβ [10].

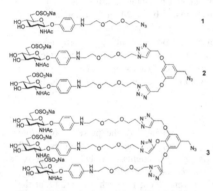

Figure 1 Chemical structure of multivalent saccharides.

RESULTS
Preparation and Characterization of SAM

Substrates with mono-, di-, and tri -valent 6S-GlcNAc were synthesized (Figure 1). To avoid reaction complexities due to the hydroxyl groups of the sugar, we synthesized and immobilized the compounds via click reaction. Immobilization of the multivalent 6S-GlcNAc was confirmed by water contact angle goniometry, X-ray photoelectron spectroscopy, FTIR-RAS, and ellipsometry. The SAMs involving well-defined glyco-clusters on the substrate were regarded as a cell membrane mimic and used to determine how the multivalency of sugars contributes to amyloidosis of Aβ.

AFM Observation

The morphology of Aβ aggregates on the substrate was observed by D-AFM. The shapes and the sizes of Aβ on the substrates after preparation are summarized in Figure 2. Interestingly, the morphology of Aβ was dependent on the type of glyco-cluster. When Aβ was incubated on monovalent sugar, fibril formation was observed; the fibrils were 4–6 nm high, 8–12 nm wide, and 1–2 µm long. When incubated on divalent sugar, the morphology of Aβ was changed; both fibrils and globular aggregates (7–9 nm high and 200–500 nm in diameter) were observed (Figure 2). Incubation on trivalent sugar dramatically changed the morphology of Aβ aggregate to an exclusively globular form. The globular objects were 10–20 nm high and 500–600 nm in diameter.

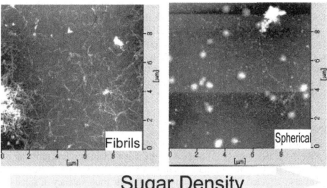

Figure 2 The morphologies of Aβ(1-42) on multivalent sugar substrate. Aβ morphology on (a) monovalent saccharides (**1**) and (b) trivalent saccharides (**3**).

Interaction Analysis with SPR

The SPR measurements were carried out to investigate the initial Aβ binding to sugars and the aggregation process. The interaction was monitored by the change of response unit (RU) on addition of 10 μM Aβ. The RU change and indicates responses of Aβ (1−42) in the order **2**(913 RU) > **3** (658 RU) » **1** (388 RU) in the measurements, hence a nonlinear RU change with valency. The high level of interaction of **2** and **3** suggested the importance of glyco-clusters.

Cytotoxicity Assay

The cytotoxicity of Aβ (1-42) to HeLa cells was evaluated by MTT assay on the substrate, where Hela cells were reported to be sensitive to the amyloid morphology. The cell viabilities with Aβ (1−42) on **1**, **2**, and **3** were 104 ± 11%, 84 ± 25%, and 76 ± 11%, respectively.

DISCUSSION

AFM observation of the substrate suggested that Aβ had a tendency to form fibrils on the suflonated sugar with low valencies, and to form globular aggregates on substrates with higher sugar valencies. The densely packed sulfonated or acidic sugars strongly induced a morphological change in Aβ aggregates. In the present case, the well-defined nanostructure of multivalent 6S-GlcNAc encouraged Aβ to aggregate in a globular morphology, indicating the significance of sugar multivalencies in amyloidosis.

The SPR measurements indicated that the 6S-GlcNAc-immobilized surfaces recognized Aβ. The amount bound of Aβ to the substrate (RU) changed with the valency, and the results of SPR was analyzed by Langmuir isotherm an Hill equation. The equilibrium dissociation constants (K_D) on **1**, **2**, and **3** were 1.49×10^{-4}, 1.16×10^{-5}, and 1.89×10^{-5} [M], respectively. The results suggest the amplification of interaction by multivalency. Therefore, the binding mode and the affinities between sugar and Aβ were related to the morphology. The strong binding of **3** (1:1) might induce the adhesion of Aβ, formation of nucleus, and subsequent globule aggregates.

MTT assay showed that the cytotoxicity of Aβ varied due to the aggregation process. Aggregation of Aβ on **3** which formed globules gave the highest cytotoxicity, followed by**2**. By contrast, aggregation of Aβ on **1** formed fibrils and had no cytotoxicity. Lambert et al. reported that cytotoxicity of Aβ (1–42) was attributed to oligomeric, small globules of Aβ (1–42), and Hoshi et al. also reported similar results. That result agreed with our observations in terms of the high cytotoxicity of Aβ globules regardless of the difference of the size of Aβ. Moreover, the multivalency of sugar variously bound to Aβ decided the secondary structure, which gives a correlation with the pathogenesis of AD.

CONCLUSIONS
This study revealed the role of sugars in amyloidosis using a well-ordered sugar display. The morphology of Aβ (1–42) was greatly influenced by the sugar valencies, and globular objects were induced to form in the presence of multivalent sugars. The binding mode between sugar and Aβ (1–42) was also affected. The results suggested that the multivalency of sugars for the amyloidosis of Aβ had a significant effect on its morphology and aggregation at the surface of the cell membrane mimic. Cell viability assay implied that higher cytotoxicity to HeLa cells was induced by Aβ which adsorbed on the multivalent sugars.

ACKNOWLEDGMENTS
This work was supported by Grant-in-Aid for Scientific Research on Innovative Areas (20106003 (YM)) and by Grant-in-Aid for Young Scientists (A) (23685027)

REFERENCES
1. F.M. Laird, H. Cai, A. V. Savonenko, M. H. Farah, K. He, T. Melnikova, H. Wen, H.C. Chiang, G. Xu, V. E. Koliatos, D. R. Borchelt, D. L. Prince, H.K. Lee, P.C. Wong, *Neurosci*, **25**, 11693-11709(2005).
2. D. J. Selko, *Physiol. Rev.*, **81**, 741-766 (2001).
3. M. Higuchi, N. Iwata, T. C. Saido, *Biochim. Biophys. Acta.*, **1751**, 60-67(2005).
4. C.G. Glabe, *J. Biol. Chem.* **283**, 29639-29643 (2008).
5. Y. Gong, L. Chang, K. L. Viola, P. N. Lacor, M. P. Lambert, C. E. Finch, G. A. Kraft, W. L. Klein, *Proc. Natl. Acad. Sci. USA*, **200**, 10417-10422 (2003).
6. M. Hoshi, M. Sato, S. Matsumoto, A. Noguchi, K. Yasutake, N. Yoshida, K. Sato, *Proc. Natl. Acad. Sci. USA*, **100**, 6370-6375 (2003).
7. J. McLaurin, T. Franklin, X. Zhang, J. Deng, P. E. Praser, *Eur. J. Biochem.* **266**, 1101-1110 (1999).
8. K. Matsuzaki, C. Horikiri, *Biochemistry*, **38**, 4137-4142 (1999).
9. T. Fukuda, S. Onogi, Y. Miura, *Thin Solid Films*. **518**, 880-888 (2008).
10. S. M. Chafekar, R. Bass, W. Scheper, *Biochimica. Biophys. Acta.* **1782**, 523-531 (2008).

Mater. Res. Soc. Symp. Proc. Vol. 1498 © 2013 Materials Research Society
DOI: 10.1557/opl.2013.338

Controlling neuronal growth and connectivity via directed self-assembly of proteins

Daniel Rizzo, Ross Beighley, James D. White and Cristian Staii
Department of Physics and Astronomy, and Center for Nanoscopic Physics, Tufts University, 4 Colby Street, Medford, MA, 02155, U.S.A.

ABSTRACT

Materials that offer the ability to influence tissue regeneration are of vital importance to the field of Tissue Engineering. Because valid 3-dimensional scaffolds for nerve tissue are still in development, advances with 2-dimensional surfaces in vitro are necessary to provide a complete understanding of controlling regeneration. Here we present a method for controlling nerve cell growth on Au electrodes using Atomic Force Microscopy -aided protein assembly. After coating a gold surface in a self-assembling monolayer of alkanethiols, the Atomic Force Microscope tip can be used to remove regions of the self-assembling monolayer in order to produce well-defined patterns. If this process is then followed by submersion of the sample into a solution containing neuro-compatible proteins, they will self assemble on these exposed regions of gold, creating well-specified regions for promoted neuron growth.

INTRODUCTION

Atomic Force Microscope based Nanolithography
 Formation of self-assembling monolayers (SAMs) is an extensively studied and diversely applied chemical process [1-3]. When a metallic substrate such as Au(111) is present in a solution of alkanethiols, individual molecules covalently bond to the surface via the sulfur head group. If left for a sufficient period of time, alkanethiols assemble onto the Au(111) surface in a $(\sqrt{3}\times\sqrt{3})R30°$ arrangement [3]. The close packed nature of this assembly causes the underlying Au surface to become chemically isolated from its environment. The Atomic Force Microscope (AFM) may then be used to both image the SAM-coated Au, and selectively remove regions of the SAM to create patterns of exposed Au. This process (called nanoshaving) may be followed by a subsequent assembly of molecules (alkanethiols, proteins, etc.) onto the then exposed Au patterns [4,5].

Neuron Growth
 Neurons have three main structural components: the cell body, axons, and dendrites. The latter two appear as finger-like projections from the cell body, which are referred to collectively as neurite processes. The axon is a much longer, thinner process that relays electrical signals to neighboring neurons, while the shorter, thicker dendrites receive them. The point at which one neuron's axon forms a connection with a neighboring dendrite is known as a synapse. Proper function of the nervous system ultimately relies on the formation and sustaining of these synapses throughout the body. As such, understanding the nature of axonal growth and synapse formation is currently an area of intense scientific research [6-10]. In time, this deeper understanding of growth processes may provide the groundwork for future medical procedures that provide nerve repair for individuals who have sustained head or spinal injuries [6, 10].

Because of the complexity of a neuron's environment *in vivo*, neuron growth and behavior is often studied in low-density samples on two-dimensional (2D) substrates. It has been demonstrated previously that Poly-D-lysine (PDL) coated surfaces effectively adhere neurons without impeding axonal growth [7]. As such, PDL coated surfaces provide a 2D environment on which neurons can naturally grow and form connections. Previous work has shown that PDL may be patterned in substrates otherwise inhospitable for neuron growth in order to direct neuron placement and growth [7,8]. Here, the alkanethiol SAMs prevent neuron growth anywhere on Au samples where PDL has not been deliberately patterned.

Here we extend this work with the aim of patterning neurons on Au circuits. Four-electrode Au circuits are formed on glass using sputter techniques. PDL is then patterned on each electrode with the intention of promoting adherence of a single neuron per electrode. This technique is particularly powerful as it could be extended to combine high precision and control over the geometry of the neuronal circuit with electrical measurements of the neuronal activity. As the neurons form connections, their electrical activity may be monitored or influenced via the underlying Au circuit.

EXPERIMENTAL DETAILS

SAM Assembly and AFM Nanoshaving

Gold samples were previously made via evaporation deposition on mica, as described in ref. [7,8]. After being stored in ethanol, these samples were immersed in a 0.1mM alkanethiol solution in ethanol and allowed to sit at 8°C for 48 hours. This results in the formation of a complete SAM on the Au surface. The SAM-coated Au sample is then removed from solution and rinsed in ethanol. While mounted in an AFM fluid cell with ethanol, the sample is first imaged at low-load so as not to disrupt the SAM layer. Subsequent high-load scans may be performed in order to selectively remove regions of the SAM from the Au, causing those alkanethiols to diffuse into the surrounding ethanol, and exposing the underlying Au surface (Fig. 1a). This process, referred to as "nanoshaving" may be used to create well-defined regions of exposed Au in otherwise SAM-coated Au samples. AFM lithography and topographical imaging studies were carried out using an an Asylum Research MFP3D AFM (Asylum Research, Santa Barbara, CA) with an inverted Nikon Eclipse Ti optical microscope (Micro Video Instruments, Avon, MA). Each sample was mounted in an Asylum Research fluid chamber, submersed in ethanol/alkanethiol or ethanol/PDL solutions and maintained at room temperature during all experiments. Silicon Nitride cantilevers (Asylum Research, Santa Barbara, CA) were used for all experiments. These cantilevers posses a nominal spring constant of 0.15 N/m.

Poly-D-Lysine Patterning

Directed assembly of PDL on Au consists of an initial step nanoshaving step, followed by an incubation period in which PDL is allowed to physisorb onto exposed Au patterns. Au samples were coated in octadecanethiol or undecanethiol SAMs according to the above procedure. Square regions of the surface with edge length of 10-20 um were chosen for nanoshaving to accommodate the size of an average neuron body. Patterns were then shaved into the SAM coated gold, roughly outlining the positioning of neuron cell bodies, as well as the axons that connect them (Fig. 1b).

a)

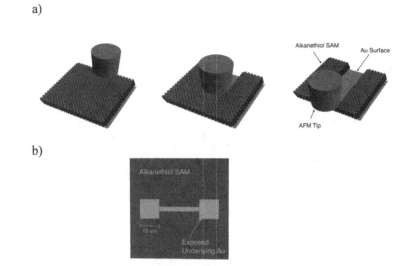

b)

Figure 1. a) Schematic of nanoshaving process b) Basic pattern for PDL directed assembly.

Once nanoshaving is complete, surrounding ethanol is replaced with an aqueous solution of PDL at a concentration of 1 mg/mL. The sample is allowed to incubate in solution for 2 hours. Because alkanethiols do not provide an effective platform for protein physisorption, PDL will only assemble on those regions of the gold that have been nanoshaved.

Plating Neurons

Embryonic rat cortices were harvested from day 18 rats at Tufts Medical School. The corticies were incubated in 5 mL of trypsin at 37°C for 20 minutes, then the trypsin was inhibited with 10 mL of neurobasal medium (Life Technologies) supplemented with GlutaMAX, b27 (Life Technologies), and pen/strep, containing 10 mg of soybean trypsin inhibitor (Life Technologies). The neurons were then mechanically dissociated, centrifuged, the supernatant removed, and the cells were resuspended in 20 mL of neurobasal medium containing L-glutamate (Sigma-Aldrich, St. Louis, MO). The neurobasal media was implemented to support neuronal growth without the use of serum, thereby reducing glial cell proliferation [9]. The cells were re-dispersed with a pipette' counted and then plated on the prepared PDL-patterned Au (which must be sterilized with UV treatment for 2 hours), at a density of 250,000 cells per 2 cm^2 area of Au. The cells are then allowed to grow for 1 week at 37°C in the incubator.

Fixing Neurons

Once the plated neurons had been allowed to incubate for several days, they were deliberately fixed in their positions on the Au substrate. Although the addition of formalin kills the neurons, it also promote the production of crosslinks between the neuron cell membrane and the surrounding substrate. The cells can therefore be easily imaged either optically or with the AFM. For fixing the neurons, first the surrounding cell media is replaced with a PBS (pH 7.2) solution 4% in formalin and allowed to sit for 30 minutes. This is then replaced with PBS, and allowed to rest for 5 minutes. This is repeated every 5 minutes 4 times.

DISCUSSION

Patterning PDL

Evaporated (transparent) 30nm Au was coated in undecanethiol SAM. Figure 2 shows an example of controlled geometry achieved via nanoshaving. The figure shows an array of 9 naoshaved squares patterned on a SAM coated Au surface. Each square has a linear dimension of 10 µm (approximately the size of a neuronal cell) and it is located approximately 10 µm apart from neighboring squares. As demonstrated in our previous work [7,8], neuron cell bodies adhere with high probability on regions with patterned PDL and with very low probability on regions not specifically patterned with PDL and thus covered with PEG (see also Fig. 3b). This approach is significant because it essentially excludes neuron adhesion in all areas other than where the inhibitory (PEG) substrate is removed.

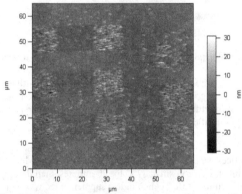

Figure 2. An example of 3 x 3 array of nanoshaved squares patterned on SAM coated Au. Each square is of $(10 \ \mu m)^2$ and positioned approximately 10 µm apart from its nearest neighbors.

Neuronal Growth on Au Circuits

We are using clean room microfabrication techniques (shadow mask evaporation, stamp patterning and dielectrophoresis) to create Au electrodes on silicon dioxide substrates. Then using AFM nanografting we create protein patterns on Au electrodes (Fig. 3a), and then culture neuronal cells on these patterns. These Au patterns can then used as microelectrodes for stimulating and recording neural activity.

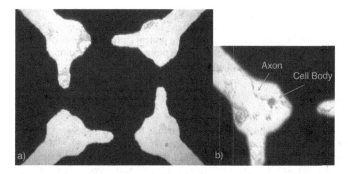

Figure 3. (a) Optical microscope image of sputtered Au circuit pattern. Four nanoshaved protein squares (~10um in size) are visible in the center of each electrode. The squares are intended to control the position of neurons. (b) Higher magnification optical image showing the adherence of a neuronal cell to the PDL pattern.

Nanoshaving allows us to have very good control over cell-cell distance, axonal elongation length and the underlying type of biochemical cues (protein type). We hypothesize that measuring the parameters related to neuronal growth on different protein patterns (rate of growth, adhesion strength, stiffness) will yield systematic and fundamental information about the role played by various growth factors in determining axon/dendrite outgrowth and connectivity. The responses may due to different chemical epitopes related to cell adhesion and signaling, more subtle differences in hydrophobicity, or to differences in secondary structure and thus topography among the different proteins. In addition, using standard microfabrication techniques we can create micrometer-size electrodes to *immobilize and electrically stimulate* neuron somata, and to *measure* the electrical activity (action potentials) in patterned neurons.

CONCLUSIONS

The success of PDL patterning on SAM-coated Au demonstrated the validity of the above-described patterning technique. It is well known that axon growth and guidance are governed by a very large parameter space, where neuronal chemotaxis, cell-cell signaling, biochemical and biomechanical cues all play an important role. The techniques described here allow us to gain new insight into neuronal growth based on just a few measurable parameters in simplified geometries in 2D. Specifically, we can systematically investigate the adhesion and real time growth of neuronal processes in the presence of *different types* of patterned extracellular matrix proteins and adhesion factors (poly-lysine, laminin, fibronectin, collagen etc). Using AFM nanoshaving we can create patterns of extracellular matrix proteins and guidance factors on growth-inhibitory polyethylene glycol on gold (PEG/Au) substrates. In future experiments we will perform the following sequence of experiments: 1) start with a series of identical PEG/Au substrates; 2) use AFM to pattern a certain type of growth promoting protein (e.g. fibronectin) on one substrate and a different type (e.g. laminin) on another substrate; the protein patterns on different substrates will have identical geometries; 3) culture neuronal cells on the patterned substrates. We will therefore create a series of samples, containing identical numbers of neurons growing on identical *geometrical* patterns, while the *type (chemistry)* of the underlying growth promoting protein is different from sample to sample. For each sample we will measure: i) rate of growth on the underlying protein pattern; ii) adhesion strength and rupture forces between the neuronal process and the protein patterns. The ultimate goal of these experiments is to quantify the role that different types of biochemical cues play in neuronal growth and development.

ACKNOWLEDGMENTS

The authors thank Elise Spedden, Min Tang-Schomer, Pramod Sing, and Sameer Sonkusale for their assistance with cell culture and Au patterning. We thank Gregory Frost and Dr. Steve Moss's laboratory at Tufts Center of Neuroscience for providing embryonic rat brain tissues under Tufts University approval animal care protocols. This work was supported by the National Science Foundation (grant CBET 1067093), Tufts Faculty Research Award (FRAC) and by the Tufts Summer Scholars (support for RB).

REFERENCES

1. J. Stetner, in *Self-assembled monolater formation of alkanethiols on gold: Growth in solution versus physical vapor deposition*, Doctoral Thesis (Graz University of Technology, 2010), pp. 25-47.
2. C. O'Dwyer, G. Gay, B. Viaria de Lesegno, and J. Weiner, Langmuir **20**, 8172-8182 (2004).
3. A. Ulman, J.E. Eilers, and N. Tillman, Langmuir **5**, 1147-1152 (1989).
4. X. Zhuang, L.E. Bartley, H.P. Babcock, R. Russel, T. Ha, D. Herschlag, and S. Chu, Science **288**, 2048–2051 (2000).
5. Song Xu and Gang-yu Liu, Langmuir **13**, 127-129 (1997).
6. K. Franze and J Guck, Rep. Prog. Phys. **73,** 094601 (2010).
7. C. Staii, C. Viesselmann, J. Ballweg, L. Shi, G-y. Liu, J.C. Williams, E.W. Dent, S. Coppersmith, and M.A. Eriksson, Biomaterials **30**, 3397-3404 (2009).
8. C. Staii, C. Viesselmann, J. Ballweg, J.C. Williams, E.W. Dent, S.N. Coppersmith, and M.A. Eriksson, Langmuir, **27**, 233–239 (2011).
9. E. Spedden, J.D. White, E.N. Naumova, D.L. Kaplan, and C. Staii, Biophys. J. **103**, 868-877 (2012).
10. R. Beighley, E. Spedden, K. Sekeroglu, T. Atherton, M.C. Demirel, and C. Staii, App. Phys. Lett. **101**, 143701 (2012).

Mater. Res. Soc. Symp. Proc. Vol. 1498 © 2013 Materials Research Society
DOI: 10.1557/opl.2013.339

A Sarcomere-Mimetic Gel: Gelation of Astral-Shaped Actin Filaments with Their Plus End Connected on Photopolymer Beads by Myosin Filaments

Taiji Ikawa[1], Masahito Shiozawa[1], Makoto Mouri[1], Mamiko Narita[1] and Osamu Watanabe[1]

[1]Toyota Central R&D Labs., Inc., Nagakute, Aichi, 480-1192, Japan

ABSTRACT

We have developed a method of a stepwise construction of a gel consisting of (i) astral-shaped actin filaments with their plus end connected on photo-responsive polymer beads and (ii) bipolar myosin filaments as linkers in order to mimic sarcomeric structure, the basic unit of a muscle. In the method, firstly, 4 μm diam. beads were prepared from an acrylate polymer containing azobenzene moiety by a good-solvent evaporation technique. Next, gelsolin, which servers and remains bound to the plus end of an actin filament, was adsorbed and then immobilized on the bead surface by exposure to light from blue light-emitting diodes, and then fluorescent actin filaments were mixed with the beads. Formation of star-like, astral actin filaments on the beads were observed in fluorescent microscopy. Finally, the beads with actin filaments were mixed with myosin mini filaments with ca. 1 μm in length. Dozens of the beads were observed to be assembled into a gel form in optical microscopy. After adding adenosine triphosphate to the gel solution, the gel was slowly contract up to 60% comparing with its original volume, suggesting that linker myosin filaments moved on the actin filaments toward the plus end on the beads.

INTRODUCTION

Motor proteins have a fascinating property that they bind to a cytoskeletal filament and move steadily along it [1]. They are powered by the hydrolysis of adenosine triphosphate (ATP) to adenosine diphosphate (ADP) and convert chemical energy into mechanical work. Motor proteins cause cytoskeletal filaments to slide against each other, generating the force that drives cell division, ciliary beating and muscle contraction. The muscle contraction is caused by the ATP-driven sliding of highly organized arrays of cytoskeletal actin filaments and myosin II filaments in tiny contractile units called sarcomere. The actin filaments are attached at their plus end to a Z-disc, which represents the lateral boundary of the sarcomere and functions as a nodal point in transmission of force generated within the sarcomere. The actin filaments are overlapped with bipolar assembly of myosin II at the middle of the sarcomere. Force-producing head units stick out from the side of myosin II filaments walk toward the plus ends of two sets of actin filaments of opposite orientations, causing sarcomere shortening.

In the late 1980s it has been demonstrated that actin filaments is glide on the glass slide on which the myosin head units are attached [2, 3]. Using the experimental technique, molecular-scale mechanics of the motor protein has been studied. In the 2000s, directional control of the movement of the actin filaments was achieved on the nano patterned substrate based on the nanolithographic technology [4]. The previous methods, however, the displacement is limited to one dimension and micron scale. To our knowledge the large scale muscle contraction system based on the molecular motor has not been achieved.

In this paper, we demonstrate how to emulate skeletal muscle structure to achieve the rapid and large muscle-like contraction. We have developed a method of a stepwise construction of a gel consisting of (i) astral-shaped actin filaments with their plus end connected on photopolymer beads and (ii) bipolar myosin filaments as linkers for the gelation (Figure 1). The gel structure is intended to mimic sarcomeric structure in skeletal muscle myofibril, in which the photopolymer beads play a role as Z-disc in sarcomere. The possibility to large scale and rapid displacement is examined.

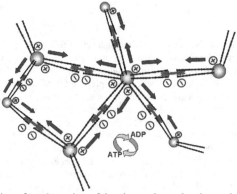

Figure 1. An illustration of a gel consists of the photopolymer beads attached with actin filaments and bipolar myosin filaments.

EXPERIMENT

Preparation of photopolymer beads

A polymer containing an azobenzene moiety (azopolymer) was employed for the material of the micro beads with Z-disc-like function. The azopolymer has the specific photo-immobilization function; the the shape of the substances on the surface is imprinted by photo-irradiation and then the substances are immobilized physically. Several different substances, e.g., polystyrene particles, DNA, Ig G, and assembled actin filaments are demonstrated to be immobilized on the azopolymer [5, 6].

In this experiment, poly{4'-[[[2-(methacryloyloxy) ethyl]ethyl] amino]-4-cyanoazobenzene-co-methyl methacrylate} (15 mol% azobenzene moiety) was obtained by free-radical polymerization (Figure 2). The polymer had a molecular weight of 25,000, a T_g of 102°C and an absorption maximum of 447 nm. Contact angle of water and ethylene glycol to the polymer were 75° and 53°, respectively.

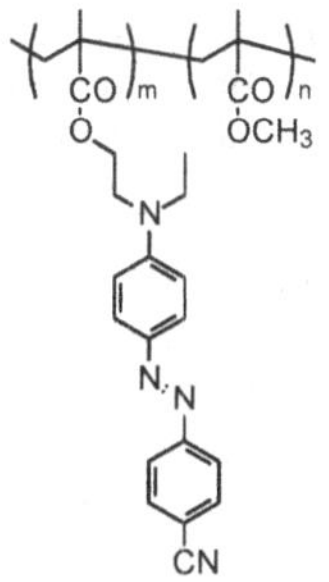

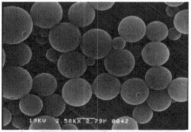

Figure 2. A chemical structure of an azopolymer.

The micro beads were prepared from the azopolymer by using a good-solvent evaporation technique shown by Yabu et al [7] . 10 mg of the azopolymer was dissolved in 100 mL of tetrahydorfuran (THF). The THF solution was dropped into the 25 mL, 50 mL and 100 mL of distilled water with being stirred vigorously by stirrer bar. After evaporation of THF, the aqueous solution was centrifuged and the formed micro beads were collected. The preparation condition and the resultant sizes of the beads were summarized in Table 1. The size of the micro beads was observed by using scanning electron microscopy (Figure 3).

Table 1. Preparation condition and resultant size of azopolymer beads.

Lot.No.	Conc. AZP (mg/mL)	THF:Water (vol:vol)	Ave. Diam. (nm)	Max. Diam. (nm)	Min. Diam. (nm)	Variation (%)
1	0.1	100:25	2770	4820	1660	19
2	0.1	100:50	740	1000	460	18
3	0.1	100:100	530	880	190	29

Figure 3. A SEM image of micro beads (Lot. 1).

Fabrication of astral shaped actin filaments on the photopolymer

Gelsolin, which servers and remains bound to the plus end of an actin filament [1], was immobilized on the azopolymer bead surface by photo irradiation and then the actin filaments were bound on the gelsolin on the beads surface.

Lyophilized gelsolin (Cytoskeleton) was dissolved in an aqueous buffer solution of 20 mM HEPES, 25 mM KCl_2, 5 mM $MgCl_2$ and 1mM Dithiothreitol (DTT) (Buffer 1) at a concentration of 400 μg/mL. The gelsolin solution was mixed with the azopolymer beads, and then the solution was irradiated with light of 10 mW/cm^2 in optical power density and 470 nm in wavelength from the array of blue light emitting diodes (LEDs) for 30 minutes. The beads was then washed 3 times using the same buffer to remove the unbound gelsolin.

Actin filament (F-actin) was polymerized from lyophilized monomeric Globular (G-) actin (Invitrogen, Alexafluor647 conjugated) dissolved in a buffer solution of 5 mM Tris HCl, 0.2mM $CaCl_2$ and 1 mM DTT (Buffer 2) at a G-actin concentration of 1 mg/mL [8]. Polymerization was initiated by adding KCl at a final concentration of 100 mM and kept intact for 30 min at room temperature. Phalloidin was added in a 1:1 M ratio with G-actin to stabilize the filament. The F-actin solution was ultracentrifuged at 50000rpm for 2 hour at 24°C. All supernatant was removed from the sample and resuspended into a Buffer 2.

The solution of 1 mg/mL F-actin and the gelsolin bounded photopolymer beads in buffer 1 containing 1mg/mL bovine serum albumin (BSA) were mixed and kept intact for 30 minutes at room temperature. The solution was centrifuged at 3000 rpm for 5 min at 4°C, and the supernatant was removed carefully without drying and resuspended in a Buffer 2. The process repeated 3 times. Finally, the beads in the solution were observed by a Nikon fluorescent microscope equipped with Hamamatsu C9100-13 EM-CCD camera.

Myosin filament

The *in vitro* assembly of the myosin was regulated by ionic strength and preparation condition [9]. When the salt concentration of solution of myosin decreases by dilution or dialysis, the myosin molecules polymerizes into filaments. The structure of the synthetic filaments formed is not fixed but varies grossly with the final ionic condition and the rate at which the salt concentration is reduced. In this experiment, myosin from rabbit skeletal muscle (22 mg/ml containing 20 mM TES, pH7.0, 0.6 M NaCl, 50 % of glycerole, Prozyme) was dialyzed by aqueous buffer solution containing 20 mM imidazole, 0.5M KCl, 0.4 mM $MgCl_2$ and then the aqueous solution was diluted into the final concentration of 20 mM imidazole, 25 mM KCl, 4 mM $MgCl_2$. The size of the myosin filament was confirmed by using a Digital Instruments Nanoscope E scanning probe microscope in contact-mode with a standard silicon cantilever (tip radius of curvature ~ 20 nm). The length of the filament was observed to be ca.1 μm (Figure 4).

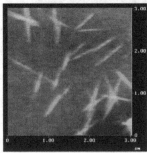

Figure 4. An atomic force microscope image of myosin filament on the substrate surface.

Gelation of astral shaped actin filaments with myosin filaments and its contraction

100 μL of the solutions of actin filament-bounded azopolymer beads and 6 μL of myosin filaments were mixed and stirred a couple of times. In this process, the beads were observed to be aggregated. The solution was penetrated into perfusion chamber (Coverwell™). The aggregation of the beads in the chamber was observed by a Nikon TE2000U optical microscope. 1μL of 100 mM ATP solution (Cosmo bio) was penetrate into the chamber and the movement of the aggregation of the beads was recorded by the CCD video camera.

DISCUSSION

The structure of the photopolymer beads were observed by fluorescent microscopy before gelation by mixing with the myosin filament. The fluorescent image is shown in Figure 5. The image clearly shows that dozens of actin filaments bound on the bead surface. During the observation the filaments showed Brownian movement, rolling heavily in the wave. The filaments, however, did not get entangled and were vertically fixed on the bead surface. This suggests that the gelsolin immobilized on the bead surface worked well to cap the plus end of the actin filament.

By mixing the photopolymer beads with myosin filaments with ca. 1 μm in length, the beads were observed to be assembled into a gel form. The gel image was shown in Figure 5 (b). After adding ATP to the gel solution, the gel was slowly contract up to 60% for 10 second from its original volume in comparison upper image with lower one in Figure 5 (b).

The gelation and contraction behavior means that the bipolar myosin filament connects the actin filaments on the different beads, and that the polarization of the actin filament on the beads is regulated so that force-generating motor domains of the bipolar myosin filaments moves on the actin filaments toward the plus end, beads surface direction, as shown in Figure 1.

CONCLUSIONS

We have shown how to emulate skeletal muscle structure consisting of the photopolymer beads as Z-disc function, actin filaments with their plus end bounded on the photopolymer beads, and bipolar synthesized myosin filaments connected on the actin filaments. After adding ATP, the muscle like structure was shown to slowly contract up to 60% comparing with its original volume, suggesting that linker myosin filaments moved on the actin filaments toward the plus end on the beads. The method will provide promising applications not only for artificial muscle, but micromechanical system and/or active optical element.

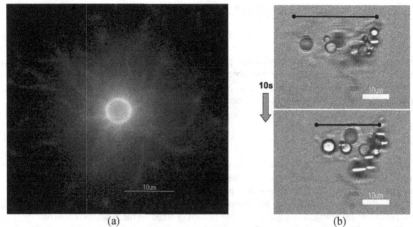

(a)   (b)

Figure 5. (a) A fluorescence image of the photopolymer beads with actin filaments. (b) The optical microscope images of the gel formed by mixing of the actin filaments-bounded photopolymer beads and myosin filaments. The gel was contract up to 60% from its original form.

ACKNOWLEDGMENTS

It is a pleasure to acknowledge many useful discussions with Dr. M. Edamatsu, Prof. Y. Y. Toyoshima and Mr. F. Hoshino. This work was supported by KAKENHI 22510121.

REFERENCES

1. B. Alberts, A. Johnson, J. Lewis, M. Raff, K. Roberts and P. Walter in Moleuclar Biology of the Cell, forth edition (Garland Science, 2002) pp.949-968.
2. S. J. Kron, J. A. Spudich, *Proc. Natl. Acad. Sci. USA.* **83**, 6272-6276 (1986).
3. Y. Y. Toyoshima, S. J. Kron. E. M. McNally, K. R. Niebling, C. Toyoshima, J. A. Spudich, *Nature,* **328**, 536-539 (1987).
4. Y. Hiratsuka, T. Tada, K. Oiwa, T. Kanayama, T. Q. P. Uyeda, *Biophys. J.,* **81**, 1555-1561 (2001).
5. T. Ikawa, F. Hoshino, T. Matsuyama, H. Takahashi, and O. Watanabe, *Langmuir* **22**, 2747 (2006).
6. T. Ikawa, Y. Kato, T. Yamada, M. Shiozawa, M. M. Mouri, F. Hoshino, O. Watanabe, M. Tawata, H. Shimoyama, *Langmuir,* **26**, 12673-12679 (2010).
7. H. Yabu, T. Higuchi, K. Ijiro and M. Shimomura, *Chaos,* **15**, 047505 (2005).
8. G. C. L. Wong, A. Lin, J. X. Tang, Y. Li, P.A. Janmey, C. R. Safinya, *Phys. Rev. Lett.,* **91** 08103 (2003).
9. G.Offer, Chapter 12 Myosin filaments, in Fibrous Protein Structure (Academic Press, 1987) pp.307-356.

Fundamentals of Assembly in Biomolecular and Biomimetic Systems

Mater. Res. Soc. Symp. Proc. Vol. 1498 © 2013 Materials Research Society
DOI: 10.1557/opl.2013.340

Effect of Modification in Cellulose Microstructure on Liquid Crystallinity

Mudrika Khandelwal[1], Nadine Hessler[2], Alan H. Windle[1]
[1]Macromolecular Materials Laboratory, Dept. of Material Science & Metallurgy, Univ. of Cambridge, UK
[2]Institute of Technical and Environmental Chemistry, Friedrich Schiller Universitat, Jena, Germany

ABSTRACT

Acid hydrolysis of cellulose leads to the formation of nanowhiskers, which can self-assemble to form a liquid crystal phase at a concentration determined by their aspect ratio. This work investigates the properties of the un-hydrolysed materials on which the length of the nanowhiskers depend. It was found that the length is determined by the branching pattern of the cellulose microfibrils and the crystallinity of the material.

INTRODUCTION

Cellulose, the most abundant biopolymer, is present in various life forms ranging from plants, trees, sea squirts, to algae and is also produced by various strains of bacteria as a exo-polysaccharide membrane. In Nature, cellulose occurs as a supra-molecular structure composed of cellulose chains forming long and thin microfibrils and ribbons. These microfibrils possess attractive mechanical properties on the nano-scale, with an estimated Young's modulus of about 140-160 GPa [1, 2]. In order to efficiently transfer these properties to macroscopic products such as fibres, these microfibrils need to be aligned.

The cellulose microfibrils are entangled, inhibiting the possibility of alignment in their natural state. In order to overcome entanglement, microfibrils are acid hydrolysed resulting into anisotropic rigid rod-like nanowhiskers, which are much shorter in length than microfibrils. On increasing the cellulose concentration, the aqueous suspensions of these nanowhiskers undergo transition from an isotropic to a liquid crystalline phase, where the nanowhiskers are locally aligned and form ordered domains. The aspect ratio, diameter, surface charge etc. of nanowhiskers determines this phase transition, which in turn are dependent on cellulose source. Bacterial cellulose yields nanowhiskers with a higher aspect ratio in comparison to the nanowhiskers obtained from plant-based cellulose [3, 4]. The effect of the hydrolysis conditions on the nanowhiskers properties such as length, diameter and aspect ratio, has been widely investigated but not enough work has been reported on correlating the effect of variation in the cellulose microstructure to the nanowhiskers obtained after their hydrolysis. In an attempt to fill this gap, nanowhiskers were obtained from bacterial cellulose of various microstructures (using various strains, additives during culture and a commercial grade bacterial cellulose nata-de-coco which is produced by fermentation of coconut milk) and their properties were correlated to the parent morphology. To our knowledge, this work is first attempt to relate microstructure of parent cellulose material with nanowhiskers obtained from hydrolysis.

MATERIALS AND METHODS

Food grade bacterial cellulose (BC) was obtained as Nata-de-Coco (NdC) from Yeguofood Co. Ltd. NdC was washed to remove preservatives and dried. BC was produced by two strains of Acetobacter Xylinum AX-DSM14666 (AX5) and ATCC 23769 (AY) (BC_AX5 and BC_AY), obtained from Dr. Dana Kralisch lab, in accordance to the method described by Hessler et. al.

[5]. In-situ modification was performed by adding 2 wt% of CMC (BC_CMC) and 0.1 wt% of calcofluor (BC_calco) at the beginning of the BC culture and the pellicles were collected at the end of 14 days. Each of the dried cellulose samples was treated with 40% (v/v) sulphuric acid (65 wt/ vol %) at 45°C for 3-4 hours. The suspensions obtained were repeatedly washed with deionised water till the pH reached close to neutral. Suspensions of various concentrations were allowed to undergo phase separation under gravity and the volume fractions of phases were measured. An Olympus microscope with polarizers was used to study transition from isotropic phase to liquid crystalline phase. The samples were coated for 60s with gold and analysed with JEOL 5800 microscope using an acceleration voltage of 5 kV and a current of 12 mA in secondary electron emission mode. Atomic force microscopy (AFM) was carried out on cellulose nanowhiskers suspensions in tapping mode on a Nanoscope SPM microscope (Veeco Instruments) with a Bruker high resolution silicon tip (also called RTESP or tap30).

RESULTS AND DISCUSSSION

Characteristics of nanowhiskers and liquid crystal phase formation

The acid attacks the easily accessible glycosidic bonds in the amorphous regions of the cellulose microfibrils, leading to the formation of rigid rod-like nanowhiskers. Fig 1(a) shows the AFM image of the nanowhiskers obtained after hydrolysis of BC. The distribution of the aspect ratios of the nanowhiskers was obtained from the AFM measurements and is depicted in the Fig. 1(b). The aspect ratio varies from below 50 to over 150. These nanowhiskers form stable aqueous suspensions, which undergo transition from isotropic to liquid crystalline phase with an increase in concentration as shown by the POM images (increase in bright regions imply increase in liquid crystalline phase volume) and the volume anisotropic phase at various concentrations in Fig. 1 (c). The onset of liquid crystal phase formation starts at about 0.05 wt% and completes at around 3 wt%, when the suspensions is completely liquid crystalline. It is evident from the literature that there is a clear dependence ratio of the transition on the aspect ratio involved in the system. The cotton nanowhiskers with the smallest aspect ratio about 7-10 undergo transition at a much higher concentrations of 15-40 wt% [3] in comparison to the NdC nanowhiskers with aspect ratio of 30-100 and the transition concentration range of 0.5 to 5 wt% [4]. This is consistent with the predictions by the Onsager theory, which states that larger the aspect ratio, lower is the transition concentration. Thus, there is a clear dependence of the liquid crystal behaviour of nanowhiskers on the source of cellulose as the properties of the source determine the aspect ratio of nanowhiskers. Thus supports the need to investigate the correlation between the variations in the cellulose and the resulting nanowhiskers.

Comparison of Nanowhiskers obtained from NdC and BC_AX5

BC_AX5 and NdC were hydrolysed to obtain nanowhiskers, shown in Fig. 2 (a, b). The nanowhiskers length distribution for each material was measured by AFM and is presented in Fig. 3 (a,b). The average length of the BC_AX5 nanowhiskers is about 2.42 ± 0.52 μm, which is almost twice the average length of the NdC nanowhiskers which is 1.13 ± 0.35 μm. The length of NdC nanowhiskers varies from below 0.5 μm to just over 2 μm while the maximum population has a length of 1-1.5 μm, on the other hand BC_AX5 nanowhiskers have length below 0.5 μm to over 3 μm with maximum population with length in the range of 2-3 μm. The diameter of the microfibrils of BC_AX5 and NdC before hydrolysis are the same as can be seen in the SEM images represented in Fig. 4(a,b). The individual microfibrils in both cases are about 20-30 nm, which aggregate to form wider ribbon like structures. This aggregation results in a branch like morphology of cellulose fibrils featured by bundling of microfibrils and splaying of ribbons.

In the case of polymers, the degree of flexibility is critical, which is expressed in terms of the kuhn length and the persistence length which determine the stiffness and straightness of the chain. Similarly, in the case of cellulose, a supra-molecular polymer, equivalent of kuhn length may be considered similar to the distance between the branching or splaying points mentioned above. The length distribution of these segments was obtained from SEM images of each material and is presented in Fig 3 (a,b) along with the length distribution of nanowhiskers obtained in each case. The average length of segment between branching points is larger for BC_AX5 (2.20 ± 0.65 μm) than that for NdC (1.13 ± 0.35 μm), which is similar to the trend observed for the length distribution of nanowhiskers. Moreover, the distribution trends of length segments and length of nanowhiskers are similar for both the material.

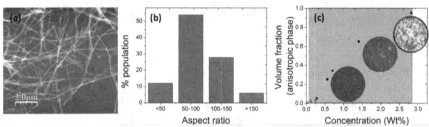

Figure 1. a) AFM image of BC_AX5 nanowhiskers b) Distribution of aspect ratio of nanowhiskers c) Phase transition diagram for nanowhisker suspension showing anisotropic phases as the birefringent regions in image

Since the nanowhiskers are obtained from the removal of amorphous region, the % crystallinity is an important parameter. XRD patterns, as shown in Fig. 8, were used to calculate the crystallinity and it was found that the crystallinity of BC_AX5 is about 90% and that of NdC is about 65% (calculated using the equation $Cr(\%)=(I_{200}-I_{am\ at18°}))/I_{200}$). As expected the length of nanowhiskers obtained decreases with decrease in crystallinity and thus the nanowhiskers length is smaller for NdC, which also has lower crystallinity than BC_AX5. The variation in microstructure between BC and NdC may arise from the differences in the cultivation conditions. This is the first evidence of the possible correlation between the morphology of cellulose and the length distribution of nanowhiskers.

Nanowhiskers obtained from BC from different bacterial strains

The microstructure of the cellulose pellicle formed varies with the strain of bacteria responsible for the synthesis. The strain AX produced a tough thick pellicle of cellulose while the other strain AY failed to produce a pellicle. Cellulose obtained from both the strains was treated with acid to obtained nanowhiskers. The AFM image and the distribution of BC_AX5 and BC_AY nanowhiskers are presented in Fig. 2 (a,c) and 3(a,c). Nanowhiskers obtained from BC_AY are much shorter with average length of 1.22 ± 0.48 μm as compared to BC_AX5 nanowhiskers with average length of 2.42 ± 0.52 μm. The SEM images of cellulose obtained from both the strains are shown in Fig. 4 (a,c) and also the distribution of length segment between branching splaying points is presented Fig. 3 (a,c). Similar to the above observation about NdC and BC, the length distribution of nanowhiskers length here is consistent with the trends of the length segment distribution in the un-hydrolysed material, with average length segment of 0.86 ± 0.12 μm and 2.20 ± 0.65 μm for BC_AY and BC_AX5 respectively. The crystallinities of the cellulose produced by two strains are 90% for BC_AX5 and 67% for BC_AY, calculated from the XRD pattern shown in Fig. 8. Again a decrease in length of nanowhiskers is seen with decrease in crystallinity.

Figure 2. AFM images of nanowhiskers from a) BC_AX5 b) NdC and c) BC_AY

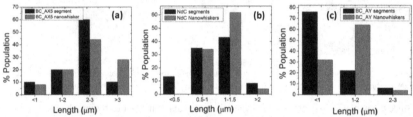

Figure 3. Distribution of microfibril length segments between branching points and the length of nanowhiskers obtained after acid hydrolysis for a) BC_AX5 b) NdC and c) BC_AY from over 100 measurements

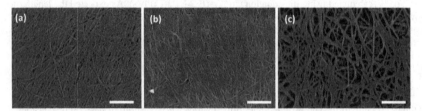

Figure 4. SEM images of a) BC_AX5 b) NdC and c) BC_AY (bar represents 1μm)

<u>Testing a proposed hypothesis with respect to nanowhiskers length control</u>

From the above two studies on the three kinds of bacterial cellulose grown with different strains and conditions, there seems to be a correlation between the morphology of the cellulose and the length of nanowhiskers. It is seen that the length of nanowhiskers is dependent on the crystallinity and the Kuhn equivalent length segment of microfibrils. On the basis of this observation, in this section, predictions are made on the nanowhiskers length by analysing the microstructure of the parent cellulose material and validated with the experimental results. In order to obtain different bacterial cellulose morphologies, culture was modified by CMC and calcofluor. CMC and Calcofluor is also known to penetrate between glucan aggregates synthesised and extruded at closely spaced sites, before they intersect and crystallise into microfibrils. Thus, the crystallinity is less in both cases in comparison to the unmodified bacterial cellulose.SEM images of unmodified bacterial cellulose, BC_CMC and BC_calco are shown in Fig. 5. The distribution of length segment between branching points is also shown in Fig. 6, and it is seen that the average length segment for BC_CMC and BC_calco are 1.93 ± 0.3 µm and 0.94 ± 0.22 µm, which are smaller than that for unmodified BC_AX5 with average length segment of 2.20 ± 0.65 µm. As far as crystallinity is concerned, the crystallinity of BC_CMC and BC_calco are 72% and 54%, which are less than that of the unmodified cellulose (90%). Therefore according to the hypothesis, on hydrolysis with acid, BC_CMC and BC_calco nanowhiskers should be smaller in length than the nanowhiskers from unmodified bacterial cellulose, with BC_calco nanowhisker being the smallest. The materials were hydrolysed to obtained nanowhiskers. From AFM images shown in Fig. 7, the average lengths of nanowhiskers in both the cases, 1.04 ± 0.36 µm and 0.86 ± 0.12 µm for BC_CMC and BC_calco respectively, are smaller than nanowhiskers from unmodified bacterial cellulose as predicted. The length distribution follows the distribution of length segments as shown in Fig. 6. The above examples prove that there is a relation between microstructure (branching pattern), crystallinity and nanowhiskers that are obtained after hydrolysis. The observations are summarised in Fig. 9 where a close relation between nanowhiskers length and branching is seen and both of them are found to increase with increase in crystallinity.

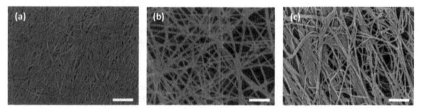

Figure 5. SEM image of a) BC_ AX5 b) BC_CMC c) BC_calco (bar represents 1 µm)

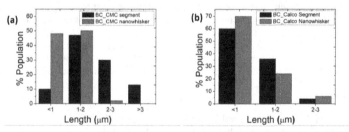

Figure 6. Distribution of microfibril length segments between branching points and the length of nanowhiskers obtained after acid hydrolysis for a) BC_CMC and b) BC_calco

Figure 7. AFM images of nanowhiskers obtained from acid hydrolysis of a) BC_ AX5 b) BC_CMC c) BC_calco

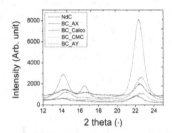

Figure 8. XRD pattern for BC_AX, BC_ AY NdC, BC_CMC, BC_calco

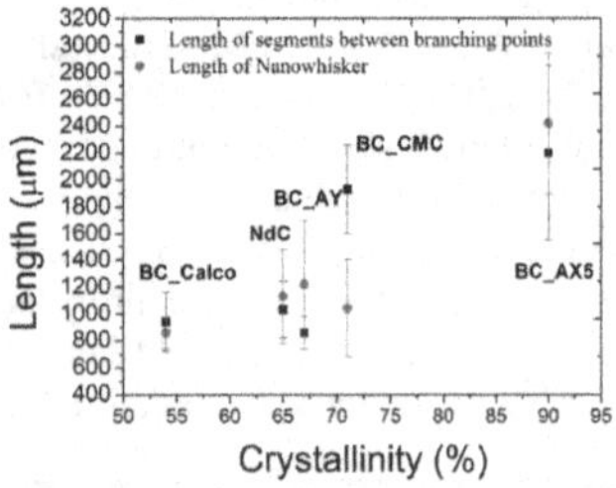

Figure 9. Variation in length between branching points and length of nanowhiskers with the % crystallinity

CONCLUSION

We have explored the relation between the microstructure of cellulose and the nanowhiskers obtained from their acid hydrolysis. The investigation was made on NdC and BC produced by two strains (AX and AY), and two additives (CMC and Calcofluor) during bacterial cell culture, which were hydrolysed to obtain nanowhiskers. The work presented here indicates that the length of nanowhiskers closely follow the length of microfibril between bundling-debundling points in the parent cellulose microstructure. It is also seen that the nanowhiskers length decreases with decrease in crystallinity of the material. The nanowhiskers length was found to decrease when both the crystallinity of the material and also the length segments between branching points decreased. Further work is needed to decouple the effect of both the parameters on the nanowhiskers length, which might be a step towards tool to control nanowhiskers length.

REFERENCES

1.   Y.C.Hsieh, H. Yano, M. Nogi, and S. Eichhorn, Cellulose, **15,** 4 (2008)
2.   G. Guhados, W. Wan, and J.L. Hutter, Langmuir, **21,**14 (2005)
3.   X.M.Dong and D.G. Gray, Langmuir, 1997. **13,** 8 (1997)
4.   A. Hirai, O. Inui, F. Horii, and M. Tsuji, Langmuir, **25,** 1 (2008)
5.   N. Heßler, and D. Klemm, Cellulose, 2009. **16,** 5 (2009).

Mater. Res. Soc. Symp. Proc. Vol. 1498 © 2013 Materials Research Society
DOI: 10.1557/opl.2013.341

Synthesis of Liposome Reinforced with Cholesterol and Application to Transmission Electron Microscopy Observation

Marina Kamogawa[1], Takuji Ube[1], Junichi Shimanuki[2], Takashi Harumoto[1], Makoto Yuasa[3], and Takashi Ishiguro[1]

[1] Department of Materials Science and Technology, Faculty of Industrial Science and Technology, Tokyo University of Science, 2641 Yamazaki, Noda, Chiba 278-8510, Japan

[2] Material Analysis Department, NISSAN ARC, LTD., 1 Natsushima-cho, Yokosuka, Kanagawa 237-0061, Japan

[3] Department of Pure and Applied Chemistry, Faculty of Science and Technology, Tokyo University of Science, 2641 Yamazaki, Noda, Chiba 278-8510, Japan

ABSTRACT

Liposome was synthesized by using mixture of dipalmitoylphosphatidylcholine and cholesterol in the ultrapure water or physiological saline. Phase transformation temperature and vibrational mode of dipalmitoylphosphatidylcholine molecule were detected by using transmission Fourier-transform infrared spectroscopy for aqueous solution, which we developed. The liposomes were fixed on an amorphous carbon mesh for ultra-high resolution transmission electron microscopy observation and stained with platinum thymidine blue. As-prepared liposomes reinforced with cholesterol were spherical in shape with size larger than 100 nm in diameter and still stable in the vacuum. Under the strong electron irradiation condition, the solution enclosed in the liposomes became unstable and then collapsed. On the other hand, the liposome synthesized in the physiological saline sometimes contains crystallized salt. As a result, the liposome shows proper strength to hold wet material in itself in a vacuum and can be used for the transmission electron microscopy observation.

INTRODUCTION

It is well known that liposome is composed of phospholipid bilayer, which is the basic structure of the cell-wall [1, 2], and is applied to the micro-capsule in the drug delivery system (DDS) [3-6] and in cosmetics [7, 8]. Since the function of biological soft matter usually depends on its three-dimensional conformation, nanoscale observation of biological soft matter such as protein and DNA has been studied intensively [9, 10]. Especially, transmission electron microscopy (TEM) offers images with sub-nanometer resolution, it is one of excellent methods for high resolution observation. There is, however, inevitable problem, i.e., the specimen should be settled in a vacuum. This means that the observation of raw biological soft matter is difficult. Therefore, the direct nanoscale observation method for biological soft matter is important technique which should be developed and, in this study, we try to use the liposome as a microcapsule for TEM observation of wet material.

EXPERIMENT

First of all, conventional liposome was synthesized by using 2, 3-di(hexadecanoyloxy)propyl 2-(trimethylazaniumyl)ethyl phosphate, i.e., dipalmitoylphosphatidylcholine (DPPC). However these liposomes have weak resistance to electron radiation. In this study, we examined reinforcement of the liposome with cholesterol. However phase diagram of mixture of DPPC solution contained cholesterol and the effect of the cholesterol have not been clear. Accordingly, transmission Fourier-transform infrared (FT-IR) absorption spectroscopy was performed to optimize the synthesis process of liposome reinforced with cholesterol. Usually transmission IR observation for aqueous solution is difficult, because water exhibits a large absorbance at IR region and the signals attenuates rapidly. Thus it is required to develop a FT-IR system for aqueous solution. We developed a special cell with an extremely short path length for water and aqueous solution to measure FT-IR spectra. The windows of the cell were made of CVD diamond. Then, we controlled the temperature (283 - 353 K).

In order to identify the IR absorption peaks, simulation was performed by using a program of Gaussian 03 for ab-initio molecular orbital calculation. In this program, B3LYP method [11, 12], which is based on Kohn-Sham density functional theory [13], was used for describing molecular orbital. Furthermore, 6-31G+ was also used for basis function [14, 15]. For the first step of calculation, we optimized the structure of DPPC molecule using molecular formula quoted from CAS-2644-64-6. Then, the vibrational modes for DPPC molecule were calculated.

Based on the phase transition temperature determined from IR absorption spectra, the liposome reinforced with cholesterol was prepared as follows: 4.405mg DPPC (NOF Co., Ltd), and 0.698mg cholesterol were firstly dissolved in 0.5ml chloroform (Wako Chemical Co., Ltd), which was gently evaporated at 328K to form films with lamellar structure composed of phospholipid bilayer by using a rotary evaporator. These films were dissolved again in the ultra-pure water or the physiological saline (Otsuka Pharmaceutical Co., Ltd). Then it was sonicated under a nitrogen gas atmosphere for 30min at 333 K, which is 10 K above phase transition temperature, and finally incubated at 328K again.

For the TEM observation, these liposome were dropped on an ultra-high-resolution carbon substrate (Okenshoji Co., Ltd) and negatively stained with platinum blue (TI blue) (Nisshin EM Co., Ltd) for 1min. The observations were performed using a JEOL, JEM2000FX electron microscope with a vacuum of 2×10^{-5} Pa. All images were taken using the computer-controlled minimum dose system mode which enabled us to catch exposure timing under drift-less condition without any radiation damage while focusing on areas of interest in the specimen.

RESULTS and DISCUSSION

Transmission FT-IR spectroscopy of DPPC water solutions contained cholesterol
Figure 1 shows a comparison in typical transmission FT-IR spectra at 333K between DPPC solutions and the one contained cholesterol, and the spectrum for pure water at 333K is also shown as the standard. There are absorption peaks corresponding to water molecules (indicated by H_2O stretching and H_2O bending), DPPC molecules (v_1^{DPPC}, v_2^{DPPC}, v_3^{DPPC} and v_4^{DPPC}) and CO_2 stretching, which is due to residual carbon dioxide gas through the IR optical path. Although optical path length of the specimen cell is fixed to be constant, intensities of absorbance for water molecule are different each other. This fact suggests that amount of water molecules incorporated in the lamella structure of DPPC depends on the phases. Then, temperature dependence of integrated intensity of absorbance for H_2O stretching is compared in Fig. 2. For the pure water, there is little temperature dependence. However the integrated intensity for the DPPC water solution decreases with temperature via a knick at around 299 K ($=T_1$), and changes from decrease to increase at 315 K ($=T_2$), and then turns abrupt step-up at

351 K ($=T_3$). These characteristic temperatures of T_1 and T_2 correspond to a transition temperature from the crystal phase (L_c) to the intermediate phase (non-rippled gel phase ($L_{\beta'}$) and/or rippled gel phase ($P_{\beta'}$)) and a transition temperature to the liquid crystal phase (L_α) [16, 17]. The temperature T_3 may possibly correspond to the decomposition or the degradation of L_α phase. Because intensity of absorbance is proportional to the amount of free water molecules contained in optical path of IR cell, its decreasing and its increasing mean the taking water molecule in the lamella structure of phospholipid bilayer and the reverse reaction respectively. On the other hand, DPPC water solution contained 30 mol% cholesterol shows different behavior as shown in Fig. 2, i.e., with increasing temperature L_c phase transforms to intermediate phase at T_4, which is almost same as T_1, then transforms to L_α phase at T_5, which is higher than T_2 for pure DPPC water solution, and an abrupt change as shown at T_3 does not appear up to 353 K. Additionally, the temperature change in absorbance for the intermediate phase and L_α phase is not so large or almost constant compared to the pure DPPC water solution. These facts mean the additive cholesterol has a role to stabilize the intermediate phase and the liquid crystal phase.

Based on our first-principle simulation by using software of Gaussian 03, the absorption peaks due to DPPC molecule v_1^{DPPC}, v_2^{DPPC}, v_3^{DPPC} and v_4^{DPPC} are respectively identified to be C-H_2 asymmetric stretching mode, C-H_2 symmetric stretching mode, vibrational mode due to polar (hydrophilic) head group, and cooperated vibrational mode due to two long hydrocarbon chains (hydrophobic group). Accordingly, temperature dependence of integrated intensity of absorbance for v_4^{DPPC} is plotted as shown in Fig. 3. The absorbance of v_4^{DPPC} for pure DPPC water solution increases with temperature, however, for the solution contained cholesterol the absorbance is almost constant up to T_5 and then gradually increases. These facts suggest that additive cholesterol suppress especially the motion of two long hydrocarbon chains (hydrophobic group) of DPPC molecule. In other words, additive cholesterol reinforces phospholipid bilayer of liposome. Therefore in this experiment, sonication temperature is chosen to be 333K above the phase transition at T_5.

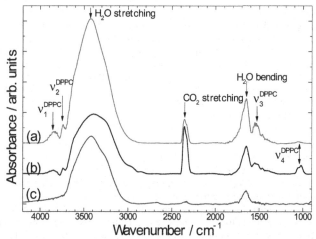

Fig. 1 Comparison in transmission FT-IR spectra at 333K of solutions. (a): DPPC water solution contained 30 mol% cholesterol, (b): DPPC water solution, and (c): pure water.

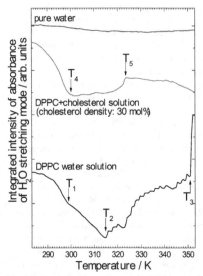

Fig. 2 Temperature change of integrated intensity of absorbance corresponding to stretching vibration for free water molecule.

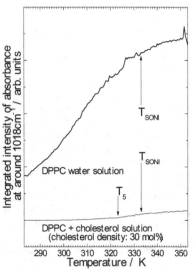

Fig. 3 Comparison in temperature dependence of integrated intensity of absorbance peak at around 1018 cm^{-1} (v_4^{DPPC}) between DPPC water solutions and the one contained cholesterol.

TEM observations

After staining with TI blue, the specimen is immediately settled in the TEM without any heating or drying. Corresponding TEM images are shown in Fig. 4. In the case of the liposome synthesized by DPPC water solution only, the contrast correspond to liposome is rear as shown in Fig. 4(a) where sheet or film-shaped images are seen, this may be burst liposome. In contrast, a lot of spherical or deformed liposomes with the size more than about 100 nm are observed as shown in Fig. 4(b). This fact suggests that the liposome reinforced with cholesterol can be used as a kind of capsule in a vacuum environment such as TEM.

In order to confirm the toughness against electron irradiation, we take photographs intermittently after strong irradiation of about 1×10^4 electrons/nm^2 on the liposome specimen with the cholesterol. To take a photograph, net number of electrons at least about 8×10^2 electrons/nm^2 is necessary. Figure 5 shows the change of the liposome. After irradiation, fluctuation on the contrast of liposome takes place from (a) to (b), then its contrast becomes blight and the inside of the liposome gets empty as shown in (c). This means that evaporation of contents is leaded by the intense electron irradiation.

When we synthesized liposome by using mixture of DPPC and cholesterol in the physiological saline, the crystallization of rock salt in the liposome is also observed as shown in Fig. 6. Selected electron diffraction pattern shows a single crystal pattern with [334] zone axis incidence. Therefore, it is deduced that nanoscaled capsule of liposome provides a stable precipitation reaction field.

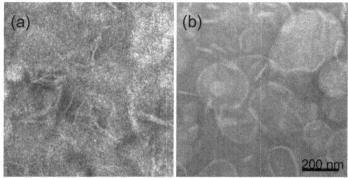

Fig. 4 TEM images of as-prepared liposomes stained using TI blue. (a): specimen prepared by using DPPC water solution, (b): specimen prepared by using DPPC water solution contained 30 mol% cholesterol.

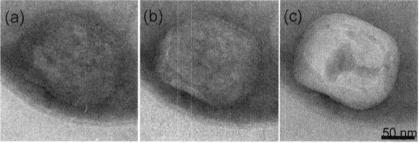

Fig. 5 Change of reinforced liposome by electron irradiation. (a): 1st photograph with exposure electron dose of about 8×10^2 electrons/nm^2, (b): 2nd photograph after strong irradiation of about 1×10^4 electrons/nm^2, and (c): 3rd photograph after additional irradiation of 1×10^4 electrons/nm^2 to the state of (b).

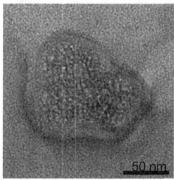

Fig. 6 TEM image of crystallized rock salt in the liposome.

CONCLUSIONS

Temperature dependence of absorbance in the water solution of mixture of DPPC and 30 mol% cholesterol were observed by using transmission FT-IR spectroscopy observation and peaks were identified based on the first-principle simulation, and then phase transition temperatures were also estimated. Accordingly, it became clear that the additive cholesterol suppress the motion of long hydrocarbon chains of DPPC molecule, and as a result reinforces phospholipid bilayer of liposome.

TEM observation suggested that the liposome reinforced with cholesterol could be used as a kind of nano-capsule in a vacuum such as TEM, provided a stable liquid-solid reaction field, although the liposome capsule was weak against intense electron irradiation.

ACKNOWLEDGMENTS

One of the authors (T. I.) was partially supported by a Grant-in-Aid for Scientific Research B (No. 23360282) from the Japan Society for the Promotion of Science, and was also supported by both a project of Center for Technologies against Cancer (2009-2013) and a project of Research Center for RNA Science (2010-2014) in Tokyo University of Science. These centers are funded by Program for Development of Strategic Research Center in Private Universities supported by the Ministry of Education, Culture, Sports, Science & Technology in Japan. M. Kamogawa would like to thank Ms. Hikaru Akaboshi, graduate student of Tokyo University of Science, who taught all of her experiences about liposome synthesis. Specimens were partially prepared by Ms. Kanako Yamamoto and Mr. Atsushi Kawashimo of Tokyo University of Science.

REFERENCES

1. A. D. Bangham, M. M. Standish, and J. C. Watkins, J. Mol. Biol. **13**, 238 (1965).
2. S. E. Schullery, C. F. Schmidt, P. Felgner, T. W. Tillack, and T. E. Thompson, Biochemistry, **19**, 3919 (1980).
3. J. A. Shabbits, G. N. C. Chiu, and L. D. Mayer, J. Controlled Release, 84, 161 (2002).
4. J. Adler-Moore and R. T. Proffitt, J. Antimicrobial Chemotherapy, **49**, Suppl. S1, 21 (2002).
5. T. Anada, Y. Takeda, Y. Honda, K. Sakurai, and O. Suzuki, Bioorganic & Medical Chemistry Lett., **19**, 4148 (2009).
6. M. Liu, L. Chen, Y. Zhao, L. Gan, D. Zhu, Wei. Xiong, Y. Lv, Z. Xu, Z. Hao, and L. Chen, Colloids and Surfaces A: Physicochem. Eng. Aspects, **395**, 131 (2012).
7. K. Arakane, K. Hayashi, N. Naito, T. Nagano, and M. Hirobe, Chem. Pharm. Bull. **43**(10), 1751 (1995).
8. K. Arakane, K. Hayashi, N. Naito, T. Nagano, and M. Hirobe, Chem. Pharm. Bull. **43**(10), 1755 (1995).
9. M. Shimizu, Y. Miwa, K. Hashimoto, and A. Goto, Bicsci. Biotech. Biochem., **57**(9), 1445 (1993).
10. J.P. Scheehana, J. M. Sheehana, E. G. Holebergb, E. E. Geisertc, and G. A. Helma, Neuroscience, **333**, 212 (2002).
11. A. D. Becke, J. Chem. Phys. **98**, 5648 (1993).
12. A. D. Becke, J. Chem. Phys. **98**, 1372 (1993).
13. W. Kohn, L. J. Sham, Phys. Review **140** (4A): A1133 (1965).
14. D. Neisius, G. Verhaegen, Chem. Phys. Lett., **66**, 358 (1979).
15. D. Neisius, G. Verhaegen, Chem. Phys. Lett., **78**, 147 (1981).
16. M. J. Ruocco and G. G. Shipley, Biochem. Biophys. Acta, **629**, 309 (1982).
17. H. J. Hinz and M. J. Sturtevant, J. Biol. Chem., **247**, 6071 (1972).

Mater. Res. Soc. Symp. Proc. Vol. 1498 © 2013 Materials Research Society
DOI: 10.1557/opl.2013.342

Design and Characterization of Nanostructured Biomaterials via the Self-assembly of Lipids

Paul Ludford, Fikret Aydin and Meenakshi Dutt

Chemical and Biochemical Engineering, Rutgers- The State University of New Jersey, Piscataway, New Jersey 08854, USA

ABSTRACT

We are interested in designing nanostructured biomaterials using nanoscopic building blocks such as functionalized nanotubes and lipid molecules. In our earlier work, we summarized the multiple control parameters which direct the equilibrium morphology of a specific class of nanostructured biomaterials. Individual lipid molecules were composed of a hydrophilic head group and two hydrophobic tails. A bare nanotube encompassed an ABA architecture, with a hydrophobic shaft (B) and two hydrophilic ends (A). We introduced hydrophilic hairs at one end of the tube to enable selective transport through the channel. The dimensions of the nanotube were set to minimize its hydrophobic mismatch with the lipid bilayer. We used a Molecular Dynamics-based mesoscopic simulation technique called Dissipative Particle Dynamics which simultaneously resolves the structure and dynamics of the nanoscopic building blocks and the hybrid aggregate. The amphiphilic lipids and functionalized nanotubes self-assembled into a stable hybrid vesicle or a bicelle in the presence of a hydrophilic solvent. We showed that the morphology of the hybrid structures was directed by factors such as the temperature, the rigidity of the lipid molecules, and the concentration of the nanotubes. Another type of hybrid nanostructured biomaterial could be multi-component lipid bilayers. In this paper, we present approaches to design hybrid nanostructured materials using multiple lipid species with different chemistries and molecular chain stiffness.

INTRODUCTION

Our goal is to design hybrid nanostructured biomaterials for use in controlled release applications such as targeted drug delivery, sensing and imaging. These applications require a material platform that can store active compounds and release them upon demand. Our earlier investigations on hybrid nanostructured biomaterials composed of functionalized nanotubes and amphiphilic lipid molecules demonstrated the formation of hybrid lipid bilayers [1 - 5.] We demonstrated that the morphology of the hybrid aggregate depended upon the concentration of the nanotubes [2], their functionalization [2], the temperature and the molecular chain stiffness of the lipid hydrophobic tails [4.] In this paper, we present our investigations in designing hybrid nanostructured biomaterials using lipid species with different chemistries and molecular chain stiffness.

Via the Dissipative Particle Dynamics (DPD) approach [1 – 6] we will investigate the mixing, phase separation and domain formation of a two-component lipid vesicle. Individual lipids are composed of a hydrophilic head group and two hydrophobic tails. We begin with a stable pre-assembled vesicle composed of two species of amphiphilic lipid molecules varying in chemistry or lipid hydrophobic tail molecular rigidity. We investigate the self-organization of the two-component system in weak, intermediate and strong segregation regimes. In addition, we will present our findings on the effect of the molecular rigidity of the lipid molecules. These investigations can potentially guide the design and creation of effective controlled-release

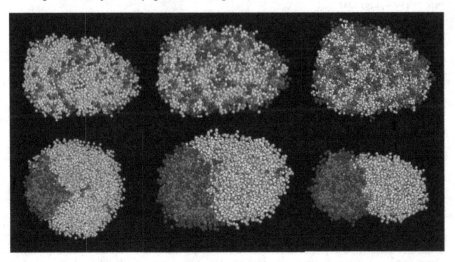

Figure 1. Snapshots of mixed (*top row*) and phase separated (*bottom row*) two component lipid bilayer vesicles with 20 % (*left*), 40 % (*middle*) and 50 % (*right*) concentration of lipid specie 2.

vehicles for lab-on-chip devices for applications in the areas of bionanotherapeutics, sensing and catalysis.

THEORY

The hydrodynamic behavior of the system can be captured by a MD-based technique known as DPD [1 – 6.] DPD is a mesoscopic particle-based simulation method that captures the molecular details of the components through soft-sphere coarse-grained (CG) models and reproduces the hydrodynamic behavior of the system over extended time scales. The DPD method is highly effective in resolving the dynamics of complex fluids over extended length and time scales due to the soft-repulsive interaction forces. The DPD technique involves integrating the dynamics of soft spheres via Newton's equation of motion through similar numerical integrators used in other MD-based simulation methods [7, 8]. The soft spheres interact via a soft-repulsive force ($\mathbf{F}_{c.ij} = a_{ij}(1 - \dfrac{r_{ij}}{r_c})\hat{\mathbf{r}}_{ij}$, for $r_{ij} < r_c$ and $\mathbf{F}_{c.ij} = 0$, for $r_{ij} \geq r_c$), a dissipative force ($\mathbf{F}_{d.ij} = -\gamma\omega^d(r_{ij})(\hat{\mathbf{r}}_{ij} \bullet \mathbf{v}_{ij})\hat{\mathbf{r}}_{ij}$) and a random force ($\mathbf{F}_{r.ij} = -\sigma\omega^r(r_{ij})\theta_{ij}\hat{\mathbf{r}}_{ij}$), where $\omega^d(r) = [w^r(r)]^2 = (1 - r)^2$(for $r < 1$), $\omega^d(r) = [w^r(r)]^2 = 0$(for $r - 1$) and $\sigma^2 = 2\gamma k_B T$. a_{ij} is the maximum repulsion between spheres i and j, $\mathbf{v}_{ij} = \mathbf{v}_i - \mathbf{v}_j$ is the relative velocity of the two spheres, $\mathbf{r}_{ij} = \mathbf{r}_i - \mathbf{r}_j$, $r_{ij} = |\mathbf{r}_i - \mathbf{r}_j|$, $\hat{\mathbf{r}}_{ij} = \mathbf{r}_{ij}/r_{ij}$, $r = r_{ij}/r_c$, $\theta_{ij}(t)$ is a randomly fluctuating variable from

Gaussian statistics, ω^d and ω^r are the separation dependent weight functions which become zero at distances greater than or equal to the cutoff r_c. The maximum repulsion a_{ij} can be related to the Flory interaction parameter χ for a given bead number density $\rho = 3$ by the following relation [6] $\chi = (0.286 \pm 0.002)(a_{ij} - a_{ii})$. Each force is pairwise additive and conserves linear and angular momentum. The random and dissipative forces are constrained by certain relations that ensure that the statistical mechanics of the system conforms to the canonical ensemble [7, 8]. The forces acting on a soft sphere i due to its interactions with neighboring soft sphere j (j $\neq$ i) which is within an interaction cutoff distance of r_c are given by $F_i \pm \sum_{j \neq i} F_{c,ij} = F_{d,ij} = F_{r,ij}$.

The lipid molecules are represented by bead-spring models with individual lipids composed of a hydrophilic head group and two hydrophobic tails. The bonds between two consecutive beads in a chain are represented by the harmonic spring potential $E_{bond} = K_{bond}((r-b)/r_c)^2$, where K_{bond} is the bond constant and b is the equilibrium bond length. The three-body stiffness potential along the lipid tails has the form $E_{angle} = K_{angle}(1 + \cos\theta)$ where θ is the angle formed by three adjacent beads. This stiffness term increases the stability and bending rigidity of the bilayers [4.]

The soft repulsive pair potential parameters for the lipid molecule beads were selected in reference to experimental systems. Measurements of the average interfacial areas per lipid and the diffusion coefficients provide the physical spatiotemporal scales for our particle-based DPD model [1 - 5.]

The soft repulsive interaction parameters between the head (h), tail (t) beads of lipid types 1 and 2 and the solvent (s) beads are assigned the following values: $a_{ss} = 25$, $a_{h1h1} = 25$, $a_{t1t1} = 25$, $a_{h2h2} = 25$, $a_{t2t2} = 25$, $a_{h1t1} = 100$, $a_{h1s} = 25$, $a_{t1s} = 100$, $a_{h2t2} = 100$, $a_{h2s} = 25$, $a_{t2s} = 100$, $a_{h1t2} = 100$ and $a_{h2t1} = 100$. The values of inter-lipid species head-head a_{h1h2} and tail-tail a_{t1t2} soft repulsive interaction parameters will be varied for the mixtures of lipids with different chemical properties.

RESULTS

We use self-assembly of amphiphilic lipids in a hydrophilic solvent to generate a two-component lipid bilayer vesicle. We randomly disperse two types of amphiphilic lipid molecules and hydrophilic solvent beads in a 30 r_c x 30 r_c x 30 r_c simulation box composed of 504 lipid molecules. We set the pair interaction potentials to effectively treat both lipid molecules as a single specie to enable the formation of a stable mixed vesicle. The unfavorable enthalpic pair interactions between the hydrophilic and hydrophobic components drives the system to minimize its free energy through the self-assembly of the lipid molecules to form a lipid bilayer vesicle (see Fig. 1 *left-top row*).) We use lipid vesicles composed of two species of amphiphilic lipid molecules for our studies on the effect of different head and tail group moeities and molecular chain stiffness.

Diverse head and tail group chemistries
We use a self-assembled mixed binary lipid vesicle as our initial configuration with 20%, 40 % and 50 % concentrations of lipid type 1 (see Fig. 1 *top row*).) The dissimilarity in the amphiphilic lipids can arise due to differences in the chemistry of the head or tail groups. The chemistry of the lipid head and tail groups is modeled effectively through a soft repulsive

interaction parameter a_{ij}. The soft repulsive interaction parameter a_{ij} will determine the degree of phase separation in the binary lipid system. We set $a_{t1t2} = 30$ and $a_{h1h2} = 25$ and observe the phase segregation in the binary lipid vesicle with 20 %, 40 % and 50 % of one of the lipid species (see Fig.1 (*bottom row*).) Our results show that a small increase in the repulsion between the tail- tail interactions between the two lipid species induces a complete phase segregation of the two components. The two lipid species phase separate to reduce the line tension between dissimilar components without disrupting the structural integrity of the binary-component vesicle. We repeated our investigations on the phase segregation morphology using a different initial configuration of the mixed binary vesicle.

We use a stable pre-assembled vesicle composed of two chemically dissimilar amphiphilic lipid molecules in a 40 r_c x 40 r_c x 40 r_c simulation box composed of 1089 lipid molecules with 544 lipid molecules of specie 1 and 545 lipids of specie 2. The two lipid species are organized into two symmetrical halves of the lipid bilayer vesicle. We investigate the self-organization of the two-component lipid system under weak, intermediate and strong segregation regimes for a range of head-head a_{h1h2} and tail-tail a_{t1t2} interactions (a_{h1h2}, a_{t1t2} = 30 - 60) between the two lipid species.

For bilayers composed of lipid species with dissimilar tail groups (identical head groups, for example 1,2-Dimyristoyl-*sn*-glycero-3-phosphocholine and 1,2-Dipalmitoyl-*sn*-glycero-3-phosphocholine), we report the stability of the two distinct phases increases with the dissimilarity between the lipid species. Fig. 2 (*left*) shows the dynamics of the phases under different segregation regimes. For the weak and intermediate segregation regimes, the line tension is minimized in the initial configuration with a single interface between the two phases while simultaneously maintaining the structural integrity of the vesicle. During the simulation, we observe some degree of mixing at the interface between the two phases. These results support our observations on phase segregation starting with an initial configuration composed of a mixed binary component vesicle. In the strong segregation regime, the line tension between the two phases is sufficiently high in comparison to the interfacial tension that the two lipids at the interface repel one another, exposing the lipid tail groups to the solvent and decoupling the two hemispheres, thereby opening up the vesicle. In the mean time, the neighboring lipid head groups shield the tail groups from exposure to the hydrophilic solvent and create two micelles composed

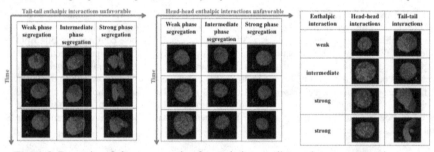

Figure 2. Dynamics of phase segregation for weak, intermediate and strong segregation regimes for different had-head (*left*) and tail-tail (*center*) interactions (*left.*) Phase diagram of phase segregation under different regimes.

of each lipid specie. Due to size constraints, the bending energy of each micelle is larger than the interfacial energy and therefore, the micelles are unable to bend and close their edges to form small vesicles.

Increasing molecular stiffness

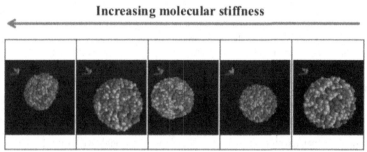

Figure 3. Phase segregation induced by different lipid hydrophobic tail stiffness for values of $K_{angle,\ 2}$ corresponding to 0, 5, 10, 15, 20 (*left* to *right*.)

For bilayers composed of lipid species with dissimilar head groups (and identical tail groups, for example 1-Palmitoyl-2-oleoylphosphatidylcholine and 1-Palmitoyl-2-oleoyl-*sn*-glycero-3-phosphoethanolamine), we find the coarsening dynamics to follow similar trends (see Fig. 2 (*middle*).) We surmise that the line tension between the distinct phases is small compared to the favorable interactions between the hydrophobic tail groups. Hence, the mixing among lipid species in the two phases with different head-head interactions but identical tail groups is limited by the lateral diffusion of the different lipid species. However, mixtures of lipid species with different tail-tail interactions have highly unfavorable enthalpic interactions in the hydrophobic region of the bilayer and consequently, a large line tension between the distinct phases. Therefore, the system will remain phase segregated to maintain the line tension at a minimum value as the free energy of the system is already at its minimum. Fig. 2 (*right*) summarizes the phase diagram of a binary-component lipid bilayer, for the weak, intermediate and strong segregation regimes.

Molecular chain stiffness
We extended our phase segregation investigations to binary mixtures comprised of lipid molecules with different hydrophobic tail rigidities (for example, saturated and unsaturated phospholipids.) We tune the molecular chain stiffness via the constant of the three-body angle potential K_{angle}. We use a stable pre-assembled vesicle composed of two amphiphilic lipid molecules with different molecular chain stiffnesses in a 40 r_c x 40 r_c x 40 r_c simulation box composed of 1089 lipid molecules with 544 lipid molecules of specie 1 and 545 lipid molecules of specie 2. The lipid species are organized symmetrically into two halves of the lipid bilayer vesicle. We set the angle potential constant $K_{angle,1}$ at 20 for lipid molecules belonging to specie 1 while the angle potential constant $K_{angle,2}$ for members of specie 2 is varied (= 0, 5, 10, 15.) Fig. 3 shows the self-organization of the two lipid components with increasing molecular chain stiffness for specie 2. For the range of $K_{angle,2}$ explored (= 0, 5, 10, 15), we did not observe any noticeable differences in the phase segregation dynamics or morphology.

CONCLUSIONS

In summary, we have shown the phase segregation in a binary lipid vesicle for different chemistries of the lipid species, and have identified segregation regimes that are conducive for preserving the structural integrity of the binary component vesicle. Our observations of the phase segregation morphology has been verified using two different initial configurations: (i) mixed binary lipid vesicle and (ii) a completely phase segregated lipid vesicle. Our investigations on the effect of molecular chain rigidity of the lipid components on the phase segregation dynamics did not lead to any conclusive findings. These investigations can potentially guide the design and creation of bio-nanostructured materials to be used as effective controlled-release vehicles.

ACKNOWLEDGMENTS

Portions of the research were conducted using high performance computational resources at the Louisiana Optical Network Initiative (http://www.loni.org), and the Rutgers Engineering Computational Cluster (http://linuxcluster.rutgers.edu/). P.L. would like to thank Derek Sturm for helpful discussions.

REFERENCES

1. M. Dutt, M.J. Nayhouse, O. Kuksenok, S.R. Little and A.C. Balazs, Current Nanoscience 7, 699-715 (2011.)
2. M. Dutt, O. Kuksenok, M.J. Nayhouse, S.R. Little and A.C. Balazs, ACS Nano 5, 4769-4782 (2011.)
3. M. Dutt, O. Kuksenok, S.R. Little and A.C. Balazs, Nanoscale 3, 240-250 (2011.)
4. M. Dutt, O. Kuksenok, S.R. Little and A.C. Balazs. Designing Tunable Bio-nanostructured Materials via Self-assembly of Amphiphilic Lipids and Functionalized Nanotubes. *MRS Spring 2012 Conference Proceedings* (in press.)
5. M. Dutt, O. Kuksenok and A.C. Balazs, *submitted.*
6. R.D. Groot and P.B. Warren, J. Chem. Phys. 107, 4423-4435 (1997.)
7. M.P. Allen and D.J. Tildesley, *Computer Simulations of Liquids* (Clarendon Press, Oxford, 2001) p. 71.
8. D. Frenkel and B. Smit, *Understanding Molecular Simulation, Second Edition: From Algorithms to Applications*, 2nd ed. (Academic Press, Cornwall, 2001), p 63.

Mater. Res. Soc. Symp. Proc. Vol. 1498 © 2013 Materials Research Society
DOI: 10.1557/opl.2013.343

Understanding Magnetite Biomineralisation:
The Effect of Short Amino Acid Sequences on the {100} and the {111} Surface

Amy E. Monnington and David J. Cooke
Department of Chemical & Biological Sciences, University of Huddersfield, Queensgate,
Huddersfield, HD1 3DH, UK

ABSTRACT

Magnetite (Fe_3O_4) formation within *Magnetospirillum magneticum* strain AMB-1 occurs under the influence of the Mms6 protein. It is hypothesised that if key iron binding sites within the C-terminus of the Mms6 protein are substituted for alanine, the protein's overall iron binding ability is diminished. In this study, an atomistic model of Mms6-driven magnetite formation was developed and the attachment of series amino acid repeats (alanine-alanine, alanine-glutamic acid & glutamic acid-glutamic acid) to the {100} & {111} magnetite surfaces were investigated. Our results suggest the substitution of glutamic acid for alanine residues significantly reduces iron binding affinity of the system, thus confirming the hypothesis. In addition, it is shown that the surface of preferable attachment is the {111} magnetite surface.

INTRODUCTION

Biomineralisation is the process by which living organisms form minerals. The earliest known example of biomineralisation is that of the biosynthesis of magnetite (Fe_3O_4) which can be traced back ~2000 Ma [1]. A major step forward in understanding the principles of magnetite biomineralisation occurred with the discovery of Magnetotactic bacteria (MTB) in 1975 [2]. These bacteria use specific proteins to aid the uptake of iron ions from the surrounding aqueous environment, and subsequently produce intracellular membrane-bound single nano-crystals of pure magnetite, called magnetosomes. They do this via a largely uncharacterised process known as biologically controlled mineralisation (BCM) [3]. The resulting magnetic nanoparticles (MNPs) align in a chain, imparting a magnetic dipole to the bacterial cell, allowing the bacteria to orient along the Earth's magnetic field lines in a process known as magnetotaxis [4].

Whilst the size and morphology of the MNPs formed is uniform, the exact composition observed is often species, or strain, dependant, suggesting an element of high biological control [5]. *Magnetospirillum* species-derived MNPs are generally of cubooctahedral shape, based on a combination of the {100} & {111} crystallographic faces. One such strain of this species is *Magnetospirillum magneticum* AMB-1, which possesses magnetosomes that are elongated along the {111} axis. Arakaki *et al.* identified a class of proteins within *M. magneticum* AMB-1 that are tightly associated with the magnetite crystal surface showing common features in their amino acid sequences [6]. Particular interest has been shown in the Mms6 protein, whose amino acid sequence is amphiphilic; possessing a membrane-spanning hydrophobic N-terminus, and a highly acidic hydrophilic C-terminal region which contains dense carboxyl and hydroxyl groups that are able to bind iron ions [7].

Numerous commercial applications for bacterial MNPs have been suggested, including the removal of radionuclides and heavy metals from waste water [8], contrast agents for MRI [9,10], magnetic antibodies [11,12], hyperthermia therapy [13] and drug and gene delivery [14,15]. However, such applications are not commercially viable at present. In order to produce

MNPs more economically the biomineralisation processes need to be further understood. Therefore, we have used molecular dynamics (based on classical atomistic potentials) to study the effect of iron binding on shortened amino acid sequence based on sequence mutations and compared them with shortened amino acid sequence version of the wild-type Mms6 sequence.

METHODS

The DL_POLY Classic code [16] was used to perform the molecular dynamic simulations using the NVT (constant number of particles, volume and temperature) ensemble, employing the Nosé-Hoover thermostat with a relaxation time of 0.5 ps. The trajectories were generated using the Verlet leapfrog algorithm [17] using a time step of 1.0 fs. The long-range Coulombic interactions were calculated using Ewald summation [18] and the short-range inter and intra-molecular interactions were described using the AMBER forcefield ff99SB [19] for the organic molecules, CLAYFF for magnetite [20] and TIP3P/Fs [21] for water.

The initial structures of the amino acid sequences were generated using the AMBERTOOLS package TLEAP and were then relaxed *in vacu*, using a combination of energy minimisation and NVT MD for 1ns at 300K in a simulation cell of dimensions 40 Å x 40 Å x 40 Å. Water was then added using the utility distributed with DL_POLY classic and run for a further 1 ns of NVT MD. Magnetite slabs, with the {100} and {111} surfaces perpendicular to the x-axis were generated using the METADISE code [22]. The relaxed amino acid sequence was then placed in the vacuum gap above the slab surface and water was added again.

The iron binding affinity of the amino acid sequences was determined by running a series of potential of mean force (PMF) simulations [23], with the distance between the centre of mass (CoM) of the amino acid sequence and the magnetite surface constrained in the x-direction to distances between ~0 Å and ~10 Å, but free to move parallel to the surface. The virial of the additional force exerted on the simulation due to this constraint was monitored. Each PMF calculation was run for 1ns and the average force was integrated with respect to the constraint distance to produce the free energy of binding.

175	180	185	190	195

...... M K S R D I E̲ S A Q S D̲ E E V E̲ L R D A L A

Figure 1. The C-terminal sequence of the Mms6 protein with the preliminarily identified key iron binding sites represented (underlined amino acids). The numbers represent that particular amino acids position in the Mms6 proteins full sequence.

Preliminary experimental results [24] indicated a number of residues within the Mms6 sequence that contribute to iron binding (Figure 1). It was hypothesised that if a carboxyl group, such as glutamic acid (E), was substituted for an alanine (A), the iron binding potential of the C-terminal sequence would reduce. Therefore, if all key iron binding sites within the C-terminal Mms6 sequence were substituted, a large reduction or total absence of iron binding ability should be observed. Based on the amino acid sequence of the residue at position 185-189 in the amino acid sequence (Figure 1), we investigated the attachment of a glutamic acid repeat group (EE), an alanine repeat group (AA) and an alanine-glutamic acid mix (AE) to the {100} and {111} magnetite surfaces.

The hypothesis that carboxylate substitution causes a reduction in iron binding is supported by an increase in free energy. Within free energy profiles, 0Å on the x-axis is an approximate representation of where the magnetite surface lies. The free energy of the simulation system as a function of amino acid sequence centre of mass (CoM) distance from the magnetite surface for AA, AE & EE attached to the {100} & {111} surfaces are shown in Figures 2, 3 & 4.

RESULTS & DISCUSSION

For the AA sequence on the {100} surface the free energy increases from a plateau of ~0 eV at ~7.75 Å distance between the magnetite surface and the amino acid sequence CoM to 3.38 eV at ~0 Å. Two small peaks are apparent at ~2 Å & ~5 Å which correspond with the distances at which there are water peaks, suggesting that the reason for the peaks is that there is a need for more energy in the system in order for the amino acid sequence to pass through those water barriers. The continued increase in free energy would suggest that more energy is needed within the system for the amino acid to be attached closer to the magnetite surface.

In AA attachment to the {111} surface, the free energy also increases from a plateau of ~0 eV, but for this surface at ~6.25 Å distance between the magnetite surface and the amino acid sequence CoM to 2.56 eV at ~0 Å. As with {100} surface, the two small peaks at ~2 Å & ~5 Å, corresponding to the water peaks, are apparent and the continued increase in free energy would suggest more energy is needed for the amino acid to be attached closer to the magnetite surface.

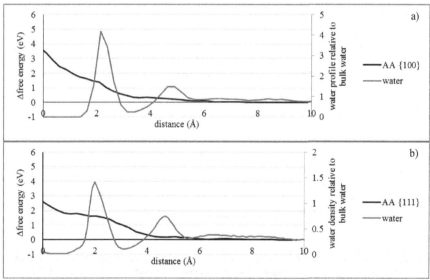

Figure 2. A comparison of the free energy profiles of an alanine-alanine (AA) sequence. a) {100} surface, b) {111} surface. Distance refers to distance between the peptide centre of mass and the magnetite surface. The grey line represents the water density profile for the system.

For AE on the {100} surface the free energy increases from a plateau of ~0 eV at ~5.75 Å distance between the magnetite surface and the amino acid sequence CoM to 5.85 eV at ~0 Å. The point at which the free energy increases to a greater extent appears to correspond with the larger of the water peaks, suggesting that the initial increase could be due to the need for more energy in the system in order for the amino acid sequence to pass through the strong water barrier. The continued increase in free energy suggests that more energy is needed within the system for the amino acid to be attached closer to the magnetite surface.

In contrast, with the {111} surface attachment the free energy decreases from a plateau of ~0 eV at ~6.5 Å distance between the magnetite surface and the amino acid sequence CoM, it then increases to 0.76 eV at ~1 Å. From that point the free energy decreases to 0.56 eV as the distance between the magnetite surface and the amino acid sequence CoM closes to ~0 Å. The distance where the free energy is at its highest corresponds with the distance at which there is a strong water peak, suggesting that the reason for the this peak could be due to need for more energy in the system in order for the amino acid sequence to pass through the strong water barrier. The energy well between ~4.5 Å & ~9 Å suggests that no energy needs to be put into the system in order for the amino acid sequence to be attached to the magnetite surface at this distance. The decrease in energy from ~1 Å down to ~0 Å suggests that the amino acid sequence more freely attaches to the magnetite surface when closer.

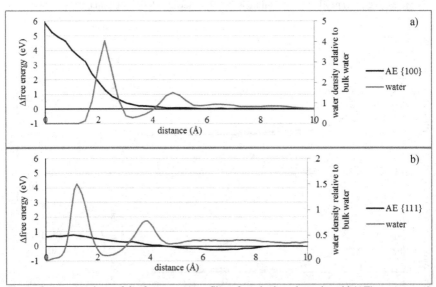

Figure 3. A comparison of the free energy profiles of an alanine-glutamic acid (AE) sequence. a) {100} surface, b) {111} surface. Distance refers to distance between the peptide centre of mass and the magnetite surface. The grey line represents the water density profile for the system.

For EE on the {100} surface the free energy increases from a plateau of ~0 eV at ~6 Å distance between the magnetite surface and the amino acid sequence CoM to 2.33 eV at ~0 Å. A small peak is apparent at ~4.6 Å which corresponds with the distance at which there is the smaller of the water peaks, suggesting that the reason for the peak is that there is a need for more energy in the system in order for the amino acid sequence to pass through the water barrier. The continued increase in free energy would suggest that more energy is needed within the system for the amino acid to be attached closer to the magnetite surface.

In contrast, with the {111} surface attachment the free energy decreases from a plateau of ~-0.03 eV at ~8.25 Å distance between the magnetite surface and the amino acid sequence CoM, it then increases to 0.42 eV at ~1 Å. From that point the free energy decreases to 0.11 eV as the distance between the magnetite surface and the amino acid sequence CoM closes to ~0 Å. The distance where the free energy is at its highest corresponds with the distance at which there is a strong water peak, suggesting that the reason for the this peak could be due to need for more energy in the system in order for the amino acid sequence to pass through the strong water barrier. The energy well between ~2.5 Å & ~8.25 Å suggests that no energy needs to be put into the system in order for the amino acid sequence to be attached to the magnetite surface at this distance. The decrease in energy from ~1 Å down to ~0 Å suggests that it is getting easier for the amino acid sequence to attach closer to the magnetite surface.

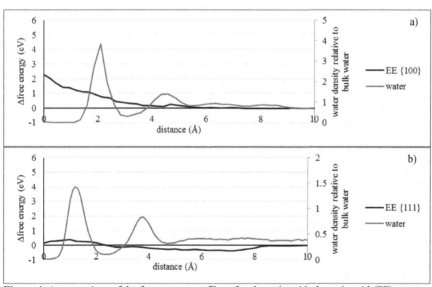

Figure 4. A comparison of the free energy profiles of a glutamic acid-glutamic acid (EE) sequence. a) {100} surface, b) {111} surface. Distance refers to distance between the peptide centre of mass and the magnetite surface. The grey line represents the water density profile for the system.

CONCLUSION

When a comparison was made between the free energy results for the different amino acid sequences on the two magnetite surfaces it was shown that, for the {100} surface, the AE sequence produces the highest free energy out of the amino acid sequences, suggesting this particular sequence has the greatest reductive effect on iron binding. Conversely, for the {111} surface, the AA repeat group produces the highest free energy out of the amino acid sequences, thus suggesting that this sequence confers iron binding inhibition. In both cases, the highest free energy was found to be at the closest distance between the magnetite surface and the amino acid sequence CoM (~0 Å). This implies that the amino acid sequence would prefer not to be attached to the magnetite surface at close range. With regard to the AE & EE sequences, the {111} surface appears to be the surface of preferable attachment as the free energy is much lower in these profiles when compared to their counterpart surfaces, this would confirm previous experimental data as this is the elongated axis within the magnetosomes and is also the axis of lowest energy (unpublished). The EE sequence appears to be the lowest energy amino acid sequence, this would confirm the experimental data as within the Mms6 sequence the residue at position 185-189 in the amino acid sequence, which contains a glutamic acid repeat, has high iron binding affinity [24] and hence should have low free energy.

This study shows that the substitution of glutamic acid, for alanine, results in an increase in free energy, thus indicating a reduction in iron binding affinity; however, whether a single or double substitution affects the iron binding more significantly is surface-dependent. In addition, it was shown that the surface of preferable attachment is the {111} magnetite surface.

ACKNOWLEDGMENTS

We acknowledge the University of Huddersfield for funding this work, the provision of computer time through the University's High Performance Computing Research Centre and its collaboration with STFC Daresbury laboratory. Additional computer time was provided via the EPSRC funded Materials Chemistry Consortium.

REFERENCES

1. J.W. Schopf, E.S. Barghoorn, D.M. Morton, R.O. Gordon, Science, **149**, 1365 (1965).
2. R. Blakemore, Science, **190**, 377 (1975).
3. D.A. Bazylinski, R.B. Frankel, Nature Reviews Microbiology, **2**, 217 (2004).
4. Y.A. Gorby, T.J. Beveridge, R.P. Blakemore, Journal of Bacteriology, **170**, 834 (1988).
5. F.C. Meldrum, S. Mann, B.R. Heywood, R.B. Frankel, D.A. Bazylinski, Proceedings of the Royal Society of London. Series B: Biological Sciences, **251**, 231 (1993); **251**, 237 (1993).
6. A. Arakaki, F. Masuda, Y. Amemiya, T. Tanaka, T. Matsunaga, Journal of colloid and interface science, **343**, 65 (2010).
7. T. Matsunaga, Y. Okamura, Trends in Microbiology, **11**, 536 (2003).
8. A.S. Bahaj, I.W. Croudace, P.A.B. James, F.D. Moeschler, P.E. Warwick, Journal of magnetism and magnetic materials, **184**, 241 (1998).

9. J.W.M Bulte, T. Douglas, S. Mann, R.B. Frankel, B.M. Moskowitz, R.A. Brooks, C.D. Baumgarner, J. Vymazal, M.P Strub, J.A. Frank, Investigative radiology, **29**, S214 (1994); Journal of Magnetic Resonance Imaging, **4**, 497 (1994).

10. C. Sun, J.S.H. Lee, M. Zhang, Advanced drug delivery reviews, **60**, 1252 (2008).

11. N. Nakamura, J.G. Burgess, K. Yagiuda, S. Kudo, T. Sakaguchi, T. Matsunaga, Analytical Chemistry, **65**, 2036 (1993).

12. N. Nakamura, T. Matsunaga, Analytica Chimica Acta, **281**, 585 (1993).

13. R. Hergt, R. Hiergeist, M. Zeisberger, D. Schüler, U. Heyen, I. Hilger, W.A. Kaiser, Journal of Magnetism and Magnetic Materials, **293**, 80 (2005); R. Hergt, S. Dutz, ibid., **311**, 187 (2007).

14. J. Dobson, Drug Development Research, **67**, 55 (2006); Gene Therapy, **13**, 283 (2006).

15. A.S. Lübbe, C. Alexiou, C. Bergemann, Journal of Surgical Research, **95**, 200 (2001).

16. W. Smith, T.R. Forester, DL_POLY is a package of molecular simulation routines, copyright The Council for the Central Laboratory of the Research Councils, Daresbury Laboratory at Daresbury, Nr. Warrington, (1996); Journal of Molecular Graphics, **14**, 136 (1996).

17. L. Verlet, Phys. Rev., **159**, 98 (1967).

18. P.P. Ewald, Annalen der Physik, **64**, 253 (1921).

19. W.D. Cornell, P. Cieplak, C.I. Bayly, I.R. Gould, K.M. Merz, D.M. Ferguson, D.C. Spellmeyer, T. Fox, J.W. Caldwell, P.A. Kollman, Journal of the American Chemical Society, **117**, 5179 (1995).

20. S. Kerisit, Geochimica et Cosmochimica Acta, **75**, 2043 (2011).

21. U.W. Schmitt, G.A. Voth, J. Chem. Phys., **111**, 9361 (2000)

22. G.W. Watson, E.T. Kelsey, N.H. de Leeuw, D.J. Harris, S.C. Parker, J. Chem. Soc., Faraday Trans., **92**, 433 (1996).

23. T.L. Hill, *Statistical Mechanics*, (McGraw-Hill, New York, 1956).

24. S.Staniland, A. Rawlings, J. Bramble, (private communication)

Directed Self-Assembly for Nanopatterning

Mater. Res. Soc. Symp. Proc. Vol. 1498 © 2013 Materials Research Society
DOI: 10.1557/opl.2013.106

Atomistic modeling of Ru nanocluster formation on graphene/Ru(0001): Thermodynamically versus kinetically directed-assembly

Y. Han[1], A. K. Engstfeld[2], C.-Z. Wang[3], L. D. Roelofs[4], R. J. Behm[2], and J. W. Evans[1,3]
[1]Department of Physics and Astronomy, Iowa State University, Ames, Iowa 50011, USA
[2]Institute of Surface Chemistry and Catalysis, Ulm University, D-89069, Ulm, Germany
[3]Ames Laboratory—USDOE, Iowa State University, Ames, Iowa 50011, USA
[4]Colgate University, Hamilton, New York, 13346, USA

ABSTRACT

Atomistic lattice-gas models for thermodynamically and kinetically directed assembly are applied to Ru nanocluster formation on a monolayer of graphene supported on Ru(0001) at 309 K. Nanocluster density, mean size, height distribution, and spatial ordering are analyzed by kinetic Monte Carlo simulations. Both models can reproduce the experimental data, but additional density functional theory analysis favors the former.

INTRODUCTION

Monolayer graphene (MLG) supported on Ru(0001) exhibits a periodically rumpled moiré structure [1-6]. The moiré structure has been attributed to the lattice mismatch between MLG and the Ru(0001) surface. A commonly accepted model corresponds to a $(12 \times 12)C/(11 \times 11)Ru$ structure [5,6]. Here, the moiré cell is a rhombus with a side length of ~ 2.98 nm, and area ~ 7.7 nm^2. Three distinct regions in the moiré cell are characterized as fcc, hcp, and atop regions refereeing to the location of the center of the C-rings relative to the Ru(0001) substrate, see Figure 1(a). The atop locations are higher by ~0.15 nm than the hcp and fcc regions. Thus, adsorption and diffusion properties of deposited metals will be also correspondingly modulated [7-11]. Experimentally observed preferential formation of Ru nanoclusters (NCs) is in the fcc region of a moiré cell (directed-assembly) [10,11], as shown in the scanning tunneling microscope (STM) image of Figure 1(b). This could reflect stronger binding or adsorption which enhances the adatom density and thus increases the nucleation rate of NC in those fcc regions (thermodynamically directed-assembly) [11].

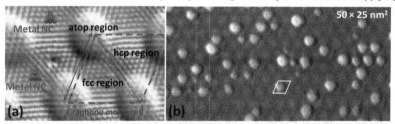

Figure 1. (Color online) (a) Schematic: Ru NCs on MLG/Ru(0001). Thick dashed blue rhombus represents a moiré cell, and thin dashed blue line is the fcc-hcp region boundary. Atop positions correspond to the vertices of the rhombus. The background is taken from STM images in [3]. (b) An experimental STM image of Ru NCs formed on MLG/Ru(0001) surface. White rhombus represents a moiré cell. Image size: 50 × 25 nm^2; temperature (T): 309 K ; flux: 0.035 ML/min; coverage (θ): 0.05 ML.

Alternatively, it could reflect a variation in diffusion barrier. A lower barrier in fcc regions would also imply a higher nucleation rate, which is proportional to the diffusion rate (kinetically directed-assembly). Can either or both models capture experiments? In reality, both the diffusion barrier and the adsorption energy will vary across the moiré cell, so the above models are idealizations. However, it is instructive to analyze these limiting cases.

Experiments not only determine the preferred location of NCs, but can also measure the filling factor (FF, the fraction of populated fcc regions), and even provide a more detailed characterization of the NC array (height distribution, size distribution, spatial correlations, etc.). In this contribution, we analyze the coverage dependence of the NC density and NC height distribution, and quantify the spatial arrangement and ordering of NCs, for both thermodynamically and kinetically directed-assembly models using kinetic Monte Carlo (KMC) simulations. Comparison is made with the experimental observations.

ATOMISTIC MODEL

Atomistic modeling requires specification of the potential energy surface (PES) describing the lateral variation of the adsorption energy for a Ru adatom on MLG/Ru(0001) surface. There are well-defined adsorption sites at the center of each carbon ring with separation $a \sim 0.25$ nm. This PES includes a short-range periodic variation with a length scale by a, and also a coarse-scale modulation on the length scale of the moiré cell by $L_M = 2.98$ nm. Figure 2(a) schematically shows the graphene moiré cell unit cell. To describe such coarse-scale variation, we will utilize a continuous periodic function [11]

$$E_p = \begin{cases} \delta \sin^2\left(\frac{\pi}{L_M}\sqrt{3}x\right) + \delta^* \sin^2\left(\frac{\pi}{L_M}y\right), & \text{for fcc region,} \\ \Delta + (\delta - \Delta)\sin^2\left(\frac{\pi}{L_M}\sqrt{3}x\right) + \delta^* \sin^2\left(\frac{\pi}{L_M}y\right), & \text{for hcp region,} \end{cases} \tag{1}$$

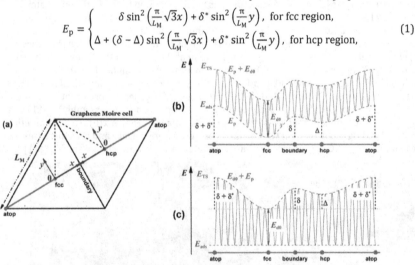

Figure 2. (Color online) (a) Schematic of the graphene moiré unit cell. (b) 1D energy cross section for thermodynamically directed assembly with varying adsorption energy. (c) 1D energy cross section for kinetically directed assembly with constant adsorption energy and varying diffusion barrier. For more details, see text.

where the coordinates x and y within the moiré cell are shown in Figure 2(a). Note that $E_p = 0$ at the center of fcc region.

For our model of *thermodynamically directed-assembly* (TDA), a one dimensional (1D) PES cross-section along the long diagonal of the moiré unit cell [indicated by the red line in Figure 1(a)] is shown in Figure 2(b). In this model, we assume that the local energy minima (E_{ads}) at the adsorption sites are always on the dashed green curve, which is described by the function E_p of Eq. (1). Also, transition state (TS) energies E_{TS} are always on the dashed red curve, which is obtained by an upward shift of E_{d0} from the E_p curve. The fine-scale variation is indicated by the oscillatory solid gray curve, see Figure 2(b). For a hop from an initial site i to an adjacent final site f, the diffusion barrier satisfies

$$E_d(i \rightarrow f) = E_{TS} - E_{ads}(i) = E_p(TS) + E_{d0} - E_p(i), \quad (2)$$

where $E_p(TS)$ is the value of E_p at the TS. We use the approximation $E_p(TS) \approx \frac{1}{2}[E_p(i) + E_p(f)]$, which immediately yields

$$E_d(i \rightarrow f) \approx E_{d0} + \frac{1}{2}[E_p(f) - E_p(i)]. \quad (3)$$

In this TDA model, Δ (δ) gives the *adsorption energy difference* between the hcp region center (the fcc-hcp boundary) and fcc region center, and δ^* gives the adsorption energy difference between the atop region and the fcc-hcp boundary.

For our model of *kinetically directed-assembly* (KDA), Figure 2(c) shows the corresponding 1D energy cross section along the long diagonal of the moiré unit cell. Now local energy minima (E_{ads}) at the adsorption sites are constant set to zero (dashed green straight line). However, the TS energy E_{TS} still varies on the coarse-scale curve (dashed red curve), which is described by E_p of Eq. (1) plus an energy parameter E_{d0}. The fine-scale variation is indicated by the oscillatory solid gray curve. Thus, for a hop from an initial site i to an adjacent final site f, the TS energy $E_{TS} \approx E_{d0} + \frac{1}{2}[E_p(i) + E_p(f)]$. Then, the diffusion barrier satisfies

$$E_d(i \rightarrow f) = E_{TS} \approx E_{d0} + \frac{1}{2}[E_p(i) + E_p(f)]. \quad (4)$$

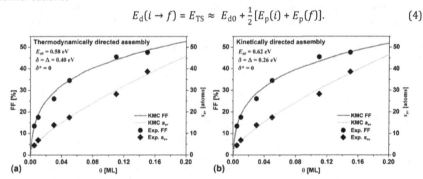

Figure 3. (Color online) Coverage dependence of filling factor, FF (in %), and NC size, s_{av} (in atoms), for Ru deposition on MLG/Ru(0001) at 309 K with fluxes ranging from ~ 0.01 to 0.1 ML/min. Experimental results are denoted by symbols, and KMC simulations by smooth curves from: (a) the TDA model, and (b) the KDA model.

Now Δ (δ) denotes the *TS energy difference* between the hcp region center and fcc region center (the fcc-hcp boundary), and δ^* between the atop region and fcc-hcp boundary.

Hopping rates of the Arrhenius form are calculated with the above E_d as

$$h(i \rightarrow f) = \nu e^{-E_d(i \rightarrow f)/(k_B T)}, \tag{5}$$

where the common prefactor $\nu = 10^{13}/s$. The complete lattice-gas model either for TDA or KDA includes random deposition on the MLG substrate, biased hopping between adjacent adsorption sites controlled by the modulated PES, and irreversible nucleation and growth of NCs. Also a "point island" model [12] is utilized for NCs, tracking only island sizes but not structures. Model behavior is assessed by KMC simulation.

We also perform first-principles DFT calculations [11] for the interaction of a Ru adatom with freestanding MLG, obtaining an adsorption energy ~ -2.0 eV and diffusion barrier ~ 0.62 eV. For Ru adatom on MLG/Ru(0001), a recent DFT calculation [13] finds an adsorption energy ~ -2.6 eV in the fcc region, but no results exist for diffusion barriers. However, it might be expected that the strong interaction of Ru with freestanding MLG is not greatly modified by underlying Ru(0001), so the diffusion barrier in the fcc region is not so different from 0.6 eV. Finally, the large bulk cohesive energy for Ru (our DFT value is -6.804 eV versus the experimental value -6.74 eV) suggests strong adatom-adatom attractions, which should produce irreversible NC formation and growth.

COMPARISON OF KMC SIMULATIONS AND EXPERIMENTS

FF analysis. FF values under experimental conditions are far below the maximum of 100%, implying facile transport of adatoms between moiré cells. KMC analysis using experimental deposition flux and temperature determines energetic parameters which can match experimental observations of the coverage dependence of the FF, as shown in Figures 3(a) and 3(b) for TDA and KDA models, respectively. For simplicity, in the TDA model, we choose $\Delta_{TDA} = \delta_{TDA} = 0.40$ eV and $\delta^*_{TDA} = 0$ to produce a strong thermodynamic driving force for NC assembly in the fcc region. The parameter E_{d0} is adjusted to match the experimental coverage dependence of the FF yielding a value of 0.58 eV, see Figure 3(a). For KDA model, for simplicity we choose $\Delta_{KDA} = \delta_{KDA}$ and $\delta^*_{KDA} = 0$. We let $E_{d0} = 0.62$ eV,

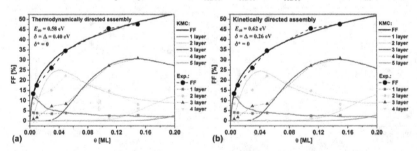

Figure 4. (Color online) Coverage dependence of NC height distribution described by the filling factors FF($h = 1, 2, \ldots$) of NCs for various specific heights h (measured in layers). Experimental results are denoted by symbols, and simulations by smooth curves from (a) TDA model, and (b) KDA model.

matching the diffusion barrier of Ru on freestanding MLG, and adjust δ to match experimental data. A good match with the experimental coverage dependence of the FF is achieved for $\Delta_{KDA} = \delta_{KDA} = 0.26$ eV, see Figure 3(b).

NC size and height distribution. The average island size is given by $s_{av} = 121\theta/$FF with s_{av} in atoms, θ in ML, and FF as a fraction. The coverage dependence of s_{av} is shown in Figures 3(a) and 3(b) for the TDA and KDA models, indicating a good match with experiment. Size distributions have the classic monomodal form expected for irreversible island formation by homogeneous nucleation on ideal (non-templated) surfaces, so coarse periodic modulation of the PES does not much affect this distribution here [11].

Our strategy [11] to elucidate the observed NC height distribution is to suppose that there exist reasonably well-defined threshold sizes (measured in number of atoms), $S_{h \rightarrow h+1}$, at which the NC makes a transition from h to $h + 1$ layers in height. Here, we also assume a hexagonal close-packed arrangement of atoms in each layer with atoms residing at 3 fold-hollow sites. We adjust these cutoffs to fit experimental observations subject to reasonable physical constraints on selected values. For hcp stacking, the minimal values are $S_{h \rightarrow h+1} = 7$, 17, 33, ... if a stable NC requires a nearest-neighbor (NN) pair in the top layer. For Ru, we find that $S_{h \rightarrow h+1} = 7, 25, 60, ...$ for $h = 1, 2, 3, ...$ layers. Thus the transition to 3 and higher layer clusters requires larger, wider clusters than the minimal pyramid. Figures 4(a) and 4(b) show the coverage dependence of NC height distribution described by the filling factors FF(h) of NCs for various specific heights $h = 1, 2, 3, ...$ layers. Experimental FF(h) results are well-reproduced by both the TDA and KDA model.

Table I. Comparison of KMC results for the TDA and KDA models for NN dimer and trimer (linear, bent, triangular) populations with experiment (Exp.). FF values are also listed.

	Coverage (ML)	FF (%)	NN dimer	Linear trimer	Bent trimer	Triangular trimer
TDA	0.01	18.87	0.0334	0.0031	0.0024	0.0021
KDA	0.01	18.71	0.0329	0.0029	0.0027	0.0021
Exp.	0.01	17.54	0.0288	0.0031	0.0018	0.0021
TDA	0.03	29.08	0.0782	0.0138	0.0126	0.0102
KDA	0.03	28.67	0.0763	0.0130	0.0119	0.0106
Exp.	0.03	26.10	0.0621	0.0090	0.0085	0.0078
TDA	0.05	34.57	0.1108	0.0258	0.0251	0.0201
KDA	0.05	34.12	0.1079	0.0234	0.0227	0.0194
Exp.	0.05	34.52	0.1096	0.0260	0.0254	0.0215
TDA	0.15	48.59	0.2216	0.0882	0.0859	0.0752
KDA	0.15	47.79	0.2142	0.0826	0.0810	0.0691
Exp.	0.15	47.62	0.2030	0.0789	0.0789	0.0697

Spatial correlations and motifs in NC array. For formation of NC's by deposition, we expect "anticlustering" in the NC array, reflecting the feature that new NCs tend to be nucleated "far" from existing NCs [11]. Here, we focus on local motifs of Ru NCs: nearest-neighbor dimers and trimers of different shapes (linear, bent, triangular). Previously, we determined experimental values for the populations of these motifs from analysis of STM images [11]. (Populations are defined so that for a random distribution of NCs, the dimer population is FF^2, and all trimer populations are FF^3.) Table I lists these populations from both TDA and KDA models as well as from experiment. There is good agreement between model and experiment, both clearly deviating from behavior for a random distribution.

SUMMARY

Thermodynamically and kinetically directed-assembly models have been applied to Ru NC formation on MLG/Ru(0001) at ~ 309 K. By analyzing filling factor, height distribution, and spatial arrangement of Ru NCs, we find that one cannot distinguish the difference between two atomistic models which both match experimental results with a suitable (non-unique) choice of energetic parameters. However, additional information is provided by expensive DFT analysis for Ru on supported MLG. Such a DFT analysis in Sutter et al. [10] does suggest that the adsorption energy is significantly higher in the fcc region than in the hcp region (and that it is fairly uniform in the latter). This favors the TDA model over the KDA model, and in fact motivated our choice of $\Delta_{TDA} = \delta_{TDA} = 0.40$ eV and $\delta^*_{TDA} = 0$. However, again we note that both these models are idealizations of actual behavior.

ACKNOWLEDGMENTS

Y.H. and J.W.E. were supported for the modeling by NSF Grant CHE-1111500. AE is grateful for a fellowship from the Fonds National de la Recherche Luxembourg (PHD09-13). A.E. and R.J.B. were supported for the experimental work by the Baden-Württemburg-Stiftung via the Competence Network "Functional Nanostructures" and the DFG via the Research Group 1376 (Be 1201/18-1). C.-Z.W. was supported for DFT analysis by the Division of Materials Sciences—BES, USDOE. Ames Laboratory is operated for the USDOE by Iowa State University under Contract No. DE-AC02-07CH11358.

REFERENCES

1. S. Marchini, S. Günther, and J. Wintterlin, Phys. Rev. B 76, 075429 (2007).
2. Y. Pan, D.-X, Shi, and H.-J. Gao, Chin. Phys. 16, 3151 (2007).
3. A. L. Vázquez de Parga, F. Calleja, B. Borca, M. C. G. Passeggi, J. J. Hinarejos, F. Guinea, and R. Miranda, Phys. Rev. Lett. 100, 056807 (2008).
4. B. Borca, S. Barja, M. Garnica, M. Minniti, A. Politano, J. M. Rodriguez-García, J. J. Hinarejos, D. Farías, A. L. Vázquez de Parga, and R. Miranda, New J. Phys. 12, 093018 (2010).
5. W. Moritz, B. Wang, M.-L. Bocquet, T. Brugger, T. Greber, J. Wintterlin, and S. Günther, Phys. Rev. Lett. 104, 136102 (2010).
6. B. Wang, S. Günther, J. Wintterlin, and M.-L. Bocquet, New J. Phys. 12, 043041 (2010).
7. Y. Pan, M. Gao, L. Huang, F. Liu, and H.-J. Gao, Appl. Phys. Lett. 95, 093106 (2009).
8. K. Donner and P. Jakob, J. Chem. Phys. 131, 164701 (2009).
9. Z. Zhao, F. Gao, and D.W. Goodman, Surf. Sci. 604, L31 (2010).
10. E. Sutter, P. Albrecht, B. Wang, M.-L. Bocquet, L. Wu, Y. Zhu, and P. Sutter, Surf. Sci. 605, 1676 (2011).
11. A. K. Engstfeld, H. E. Hoster, R. J. Behm, L. D. Roelofs, X. Liu, C.-Z. Wang, Y. Han, and J. W. Evans, Phys. Rev. B 86, 085442 (2012).
12. J. W. Evans, P. A. Thiel, and M. C. Bartelt, Surf. Sci. Rep. 61, 1 (2006).
13. B. Wang, M.-L. Bocquet, J. Phys. Chem. Lett. 2, 2341 (2011).

Mater. Res. Soc. Symp. Proc. Vol. 1498 © 2013 Materials Research Society
DOI: 10.1557/opl.2013.344

High Density Metal Oxide (ZnO) Nanopatterned Platforms for Electronic Applications

Vignesh Suresh[1], Meiyu Stella Huang[1], Madapusi.P.Srinivasan[*, 1] and Sivashankar Krishnamoorthy[*, 2]

[1]Department of Chemical and Biomolecular Engineering, National University of Singapore, Blk E5, 4 Engineering Drive 4, Singapore 117576
[2] Patterning and Fabrication group, Institute of Materials Research and Engineering (IMRE), Agency for Science Technology and Research (A*STAR), 3, Research Link, Singapore 117602

ABSTRACT

Fabrication methodologies with high precision and tenability for nanostructures of metal and metal oxides are widely explored for engineering devices such as solar cells, sensors, non-volatile memories (NVM) etc. In this direction, metal and metal oxide nanopatterned arrays are the state-of-the-art platforms upon which the device structures are built where the tunable orderly arrangement of the nanostructures enhances the device performance. We describe here a coalition of fabrication protocols that employ block copolymer self-assembly and nanoimprint lithography (NIL) to obtain metal oxide nanopatterns with sub-100 nm spatial resolution. The protocols are easily scalable down to sub-50 nm and below.

Nanopatterned arrays of ZnO created by using NIL assisted templates through area selective atomic layer deposition (ALD) and radio frequency (RF) sputtering find application in NVM and photovoltaics. We have employed NIL that produced nanoporous polymer templates using Si molds derived from block copolymer lithography (BCL) for pattern transfer into ZnO. The resulting ZnO nanoarrays were highly dense (8.6 x 10^9 nanofeatures per cm^2) exhibiting periodic feature to feature spacing and width that replicated the geometric attributes of the template. Such nanopatterns find application in NVM, where a change in the density and periodicity of the arrays influences the charge storage characteristics. The above assembly and patterning protocols were employed to fabricate metal-oxide-semiconductor (MOS) capacitor devices for investigating application in NVM. Patterned ZnO nanoarrays were used as charge storage centres for the MOS capacitor devices. Preliminary results upon investigating the flash memory performance showed good flat-band voltage hysteresis window at a relatively low operating voltage due to high charge trap density.

INTRODUCTION

Fabrication of ZnO nanopatterns with a control over geometric attributes such as feature size, spacing, dimensionality, and feature alignment with spatial resolutions below 100 nm using low-cost and high-throughput approaches are widely sought after. Typical popular techniques to prepare high feature densities ZnO over arbitrarily large areas of surface are chemically seeded growth by hydrothermal [1] or vapour-liquid-solid (VLS) [2] approaches. However, they have issues concerning lack of control over feature densities, geometric attributes and batch to batch reproducibility. Patterning of ZnO using templates obtained by techniques such as anodic aluminum oxide (AAO) [3-4], nanosphere lithography [5], colloidal template assisted patterning [6-7], shadow mask [3, 8], or NIL [9-10] processing provide an effective solution in this direction. The templates then guide the patterning of ZnO fabricated through hydrothermal [11-12], MOCVD [13] or electrodeposition of zinc [14], and direct patterning of sol-gel precursors [9]. Patterning of well-defined ZnO nanoarrays by lithography with sub-100 nm spatial resolutions that combine multiple advantages of low-cost, high-throughput fabrication and compatibility with silicon technology have not been demonstrated before.

We report here, novel process protocols based on diblock copolymer lithography (BCL) and nanoimprint lithography (NIL) as means of generating uniform nanoarrays of ZnO that could serve as a platform for a variety of applications, and for NVM, in particular. Being a high-throughput and repeatable tool, the benefit offered by the use of NIL coupled with the use of block copolymer self-assembly that is highly tunable, enables miniaturization and provides a cost effective route to fabricate high-density ZnO arrays. The processing can be performed at low temperatures, thereby making it suitable for patterning on flexible substrates for low-cost printed electronics applications.

Of late, there has been an increasing interest towards exploiting the ZnO nanostructures for fabricating electronic devices compatible with semiconductor technology. ZnO nanostructures of different qualities and morphologies offer interesting opportunities to investigate and study various material properties of ZnO in combination with the substrate properties. Semiconducting nanostructures of ZnO on transparent conductive oxides find application in solid state lighting and in solar cells, while ZnO nanopattern on silicon is widely considered as a potential combination in advanced Si electronics like memory devices, diodes and other optoelectronic devices [15-16]. ZnO nanostructures generally have a lot of zinc interstitials and oxygen vacancies making it an n-type material by nature. The behaviour of such materials on p-Si is different from that of n-Si often leading to different electronic behaviour in diodes, schottky barriers, transistors and capacitors. There are many reports on the utilization of ZnO nanostructures in capacitor memory devices, thin film transistors, diodes etc [17-19]. Other studies on the ZnO nanofilm based capacitor devices built on n-Si reports an clockwise hysteresis of only 0.12 V at 1 MHz high frequency capacitance voltage measurements (HFCV) while for a decreasing frequency sweep, a minimal or no hysteresis is observed [20-21]. These devices fall short in terms of meeting the minimum memory window required in non-volatile memory application.

As a case study, we have studied the charge storage behaviour of ZnO nanopatterns obtained through two different fabrication methodologies by incorporating them as charge storage centres in flash memory devices.

EXPERIMENTAL

The fabrication protocols involved the use of the nanoporous polymer templates (NPTs) to guide the ZnO patterning [22-23]. The NPTs were obtained through high-throughput NIL process using Si molds derived from block copolymer lithography. In short, solution of polystyrene-b-poly(2-vinylpyridine) 114 kDa, f_{PS}~0.5 in m-xylene was spin coated on Si substrate. The reverse micelles self-organized to produce quasi hexagonal assembly of nanoscale polymeric features. These features were used as etch masks for pattern-transfer into the underlying substrate through dry plasma etching processes. This resulted in the fabrication of Si nanopillars which was then used as molds in NIL. The imprinting on PMMA was carried out in a thermal nanoimprinter at 160°C for duration of 10 min. The resulting NPTs exhibited a pore size of 65 nm to guide the deposition of ZnO. The first fabrication protocol involved the use of the NPTs as guiding templates to direct ZnO deposition through radio frequency (RF) sputtering in an unbalanced magnetron (UBM) sputtering system (Nanofilm Technologies International Pte. Ltd., Singapore) to produce ZnO nanodiscs in a typical top-down approach. A 100 nm thick ZnO film deposition was sputter deposited onto the NPTs followed by ion beam milling for 250 s to remove the oxide overlayer. An acetone wash, resulted in the formation of well isolated ZnO nanopatterns of feature diameter of ~65 nm and height of ~25 nm. In the second approach, which is a bottom-up process, an atomic layer deposition (TFS200, Beneq, Vantaa, Finland) of 100 cycles of ZnO at 70°C over the nanoporous templates was carried out. This was followed by the template lift off by ultrasonicating in boiling acetone for an hour. This resulted in the ZnO nanopatterns with the feature diameter of ~55 nm and height of ~13 nm. The ZnO nanoarrays obtained were then tested for their charge storage capacitance-voltage characteristics performed using a precision LCR meter (HP4284A, Hewlett-Packard, USA) under a measurement frequency of 100 kHz.

RESULTS AND DISCUSSION

Fabrication methodologies

The AFM and FESEM images of the resulting ZnO nanoarrays obtained from two different methodologies are shown in Figure 1. As it could be observed, the two protocols employed the same Si mold for imprinting the polymer to obtain NPTs to guide the patterning and hence the periodicity (80 nm) of the resulting ZnO nanoarrays remain the same exhibiting a sub-100 nm spatial resolution.

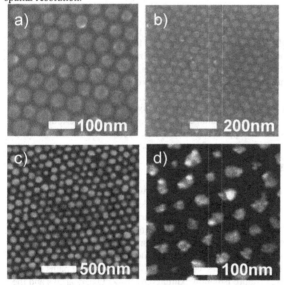

Figure 1. (a, b) FESEM images and (c, d) Tapping mode AFM of ZnO nanoarrays obtained through BCL-NIL assisted (a, c) RF sputtering and (b,d) area selective ALD.

The geometric attributes of the resulting ZnO nanopattern obtained from the top-down and bottom-up approaches are tabulated below.

Table 1. Geometric characteristics of the ZnO nanoarrays fabricated through the two different approaches.

Fabrication style	Approach	Spatial resolution	Diameter of the features (nm)	Height of the features (nm)	Feature density (per μm^2)
BCL-NIL assisted RF sputtering	Top-down	sub-100 nm	65	25	86
BCL-NIL assisted ALD	Bottom-up	sub-100 nm	55	13	86

The main advantage of the top-down assisted ZnO patterning is the use of RF sputtering for ZnO deposition onto the NPTs and ion beam milling for material removal both of which are independent of the material being deposited and hence can readily be extended to other metal oxides such as TiO_2, Al_2O_3. Thus, the fabrication process is generic in nature. The second protocol being a bottom-up process derives its advantages from the use of a PMMA polymer template, which, being passive to ALD, selectively guides the ZnO to deposit onto the exposed Si surface. This methodology too can be extended to nanopattern a range of desired materials in a more robust way. Since, the deposition can be carried out at lower temperatures, possibility of material patterning on flexible substrates widens the scope for potential integration with plastic electronics.

Thus, the protocols serve as means of producing ZnO nanoarrays of different material quality, robustness, and dimensions, and offer advantages in terms of tuning their periodicity, density thereby enabling miniaturization for application in NVM devices.

MOS capacitor device fabrication

ZnO nanopatterns fabricated through different methodologies were investigated for their electronic memory effects on the p-Si and n-Si substrates. A reliable capacitor based flash memory device is expected to offer high charge retention, high charge storage density and minimal charge leakage. With ZnO being a potential candidate to store charges and with the spatial isolation achieved through patterning, the ZnO nanoarrays are expected to be well separated from each other resulting in minimal lateral charge conduction. This would result in a device that is more resilient to stress induced leakage current (SILC) and to leakage paths created by pinhole defects. The clear feature to feature isolation exhibited by the ZnO nanoarrays and the discrete nature of the features were confirmed by performing a cross section TEM (Figure 2) of the ZnO nanoarrays.

The schematic of the capacitor test structure incorporating the ZnO nanofeatures is shown in Figure 2. A thermally grown oxide (3 nm) served as the tunneling layer and SiO_2 was sputter deposited to serve as the control oxide. The Capacitance-Voltage measurements (Figure 3) were performed at a 100 kHz frequency by sweeping the applied gate voltage from the inversion region to the accumulation region and back at room temperature with the back electrode grounded. The charge storing capacitive memory behavior of the ZnO nanopatterned arrays built on p-Si [22-23] exhibited a counter clockwise hysteresis trapping positively charged holes injected from the substrate through the tunneling oxide into the nanoarrays.

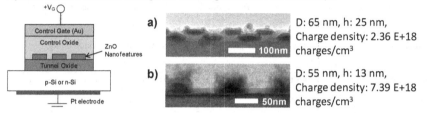

a) D: 65 nm, h: 25 nm, Charge density: 2.36 E+18 charges/cm³

b) D: 55 nm, h: 13 nm, Charge density: 7.39 E+18 charges/cm³

Figure 2. Schematic of the test capacitor device structure incorporating the ZnO nanoarrays produced from one of the two methodologies described above. Cross section TEM of the ZnO nanoarrays fabricated through BCL-NIL assisted (a) RF sputtering, (b) ALD. The diameter (D), height (h), and charge density for each of the cases is indicated.

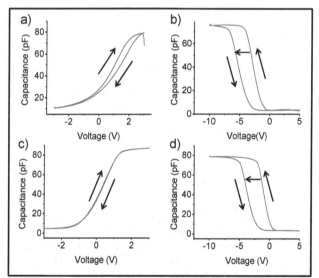

Figure 3. Capacitance-voltage characteristics of the capacitor devices incorporating ZnO nanoarrays fabricated through BCL-NIL assisted (a), (b) RF sputtering; (c), (d) ALD. (a) and (c) correspond to device structures built on n-Si while (b) and (d) employed p-Si.

The charge storage density was estimated based on the area of the hysteresis window normalized with the device dimensions and is listed in Table 2. In contrast to the devices constructed on p-type Si, the devices on n-Si, showed no hysteresis. This indicates that no electron injection into the ZnO nanopatterns took place when n-Si was used as substrate. This is likely due to the intrinsic n-type nature of the ZnO nanopatterns [24] that prevent the negatively charged electrons from being injected from n-Si and stored in the ZnO traps during the voltage sweep. Moreover, any charges stored in the interface traps or in border traps are rapidly discharged during erasure so that a negligible or no flat band voltage shift is observed.[20] The hysteresis behavior is similar to earlier reports on the ZnO capacitor structures fabricated on n-Si [20-21].

Table 2. Summary of the electrical characteristics of the ZnO nanoarrays fabricated using various techniques.

Fabrication style	Flatband voltage shift		Operating voltage		Charge storage density (per cm^3)	
	n-Si	p-Si	n-Si	p-Si	n-Si	p-Si
BCL-NIL assisted RF sputtering	Negligible shift (due to interface traps)	2.53 V	-3 V to 3 V	5 V to -10 V	----	2.36E+18
BCL-NIL assisted ALD	No shift	2.82 V	-3 V to 3 V	5 V to -10 V	----	7.39E+18

When compared to earlier reports on the electrical characteristics of ZnO nanocrystals embedded within a polyimide (PI) composite [17] (30nmC60/PI-ZnO/30nmC60) demonstrating a V_{FB} shift of 5.8V at 20V programming [25] and a V_{FB} shift of 6V for a sputtered ZnO film (200nmMSQ/50nmZnO/120nmMSQ stack) at 10V programming [16], our devices employing patterned ZnO nanofeatures as the charge storage layer exhibit superior electronic properties. The electrical characteristics of the devices built on p-Si and n-Si are summarized in Table 2.

CONCLUSIONS

We have reported novel ways to produce ZnO nanoarrays through both top-down and bottom-up approaches in the sub-100 nm regime. High-density arrays of ZnO nanostructures with different material quality, excellent pattern definition offering tunability in feature width, height and pitch using BCL-NIL assisted patterning are achieved through simple and facile methodologies. The processes are highly generic and cost effective owing to the combination of NIL and BCL techniques while at the same time offer high-resolution ZnO nanopatterns that can be extended to a range of other materials. This also provides the platform for building material components in any target applications that require patterned ZnO. Preliminary results of the electrical characteristics of the ZnO arrays show good charge storage properties for potential application in charge trap flash memories. At the same time, appropriate combinations of material quality and dimensions which our protocols are capable to deliver open up new avenues to explore such as photovoltaic (e.g. solar cells, photocatalysis), and solid state lighting.

REFERENCES

1.   L. Vayssieres, *Adv. Mater.* **15** (5), 464-466 (2003).
2.   D. Ito, M. L. Jespersen and J. E. Hutchison, *ACS Nano* **2** (10), 2001-2006 (2008).
3.   H. J. Fan, W. Lee, R. Hauschild, M. Alexe, G. Le Rhun, R. Scholz, A. Dadgar, K. Nielsch, H. Kalt, A. Krost, M. Zacharias and U. Gösele, *Small* **2** (4), 561-568 (2006).
4.   W. Lee, M. Alexe, K. Nielsch and U. Gösele, *Chem. Mater.* **17** (13), 3325-3327 (2005).
5.   WangWang, C. J. Summers and Z. L. Wang, *Nano Lett.* **4** (3), 423-426 (2004).
6.   Y. Li, N. Koshizaki and W. Cai, *Coord. Chem. Rev.* **255** (3–4), 357-373 (2011).
7.   Y. Li, W. Cai and G. Duan, *Chem. Mater.* **20** (3), 615-624 (2007).
8.   H. Chik, J. Liang, S. G. Cloutier, N. Kouklin and J. M. Xu, *Appl. Phys. Lett.* **84** (17), 3376-3378 (2004).
9.   K. Y. Yang, K. M. Yoon, K. W. Choi and H. Lee, *Microelectron. Eng.* **86** (11), 2228-2231 (2009).
10. M.-H. Jung and H. Lee, *Nanoscale Research Letters* **6** (1), 159 (2011).
11. T. U. Kim, J. A. Kim, S. M. Pawar, J. H. Moon and J. H. Kim, *Cryst. Growth Des.* **10** (10), 4256-4261 (2010).
12. H. L. Zhou, A. Chen, L. K. Jian, K. F. Ooi, G. K. L. Goh, K. Y. Zang and S. J. Chua, *J. Cryst. Growth* **310** (15), 3626-3629 (2008).
13. S. W. Kim, M. Ueda, T. Kotani and S. Fujita, *Jpn. J. Appl. Phys., Part 2* **42** (6 A), L568-L571 (2003).
14. L. Li, S. Pan, X. Dou, Y. Zhu, X. Huang, Y. Yang, G. Li and L. Zhang, *J. Phys. Chem. C* **111** (20), 7288-7291 (2007).
15. L. Schmidt-Mende and J. L. MacManus-Driscoll, *Mater. Today* **10**, 40-48 (2007).
16. N. T. Salim, K. C. Aw, W. Gao and B. E. Wright, *Thin Solid Films* **518** (1), 362-365 (2009).
17. J. H. Jung, J. Y. Jin, I. Lee, T. W. Kim, H. G. Roh and Y. H. Kim, *Appl. Phys. Lett.* **88** (11), 112107-112103 (2006).
18. F. Verbakel, S. C. J. Meskers and R. A. J. Janssen, *Appl. Phys. Lett.* **89** (10), 102103 (2006).
19. J. I. Sohn, S. S. Choi, S. M. Morris, J. S. Bendall, H. J. Coles, W. K. Hong, G. Jo, T. Lee and M. E. Welland, *Nano Lett.* **10** (11), 4316-4320 (2010).
20. S. K. Nandi, S. Chatterjee, S. K. Samanta, P. K. Bose and C. K. Maiti, *Bull. Mater. Sci.* **26** (7), 693-697 (2003).
21. S. Chatterjee, S. K. Nandi, S. Maikap, S. K. Samanta and C. K. Maiti, *Semicond. Sci. Technol.* **18** (2), 92 (2003).
22. V. Suresh, M. S. Huang, M. P. Srinivasan and S. Krishnamoorthy, *J. Mater. Chem.* **22** (41), 21871-21877 (2012).
23. V. Suresh, S. Meiyu Huang, S. Madapusi, C. Guan, H. J. Fan and S. Krishnamoorthy, *J. Phys. Chem. C* (2012).
24. S. Dengyuan and G. Baozeng, *J. Phys. D: Appl. Phys.* **42** (2), 025103 (2009).
25. F. Li, T. W. Kim, W. Dong and Y. H. Kim, *Thin Solid Films* **517** (14), 3916-3918 (2009).

AUTHOR INDEX

SUBJECT INDEX